Guillem Gilabert-Oriol
Biofouling and Organic Fouling

Also of interest

Ultrafiltration Membrane Cleaning Processes.
Optimization in Seawater Desalination Plants
Guillem Gilabert-Oriol, 2021
ISBN 978-3-11-071507-1, e-ISBN (PDF) 978-3-11-071514-9,
e-ISBN (EPUB) 978-3-11-071516-3

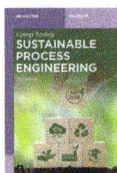

Sustainable Process Engineering
Gyorgy Szekely, 2024
ISBN 978-3-11-102815-6, e-ISBN (PDF) 978-3-11-102816-3,
e-ISBN (EPUB) 978-3-11-103032-6

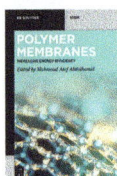

Polymer Membranes.
Increasing Energy Efficiency
Mahmoud Atef Abdulhamid, 2024
ISBN 978-3-11-079599-8, e-ISBN (PDF) 978-3-11-079603-2,
e-ISBN (EPUB) 978-3-11-079613-1

Mixed-Matrix Membranes.
Preparation Methods, Applications, Challenges and Performance
Vahid Pirouzfar, Mozhan Rostami Tehrani, Chia-Hung Su, Anahita
Hasanzad, Seyed Mohammad Hossein Hosseini, 2024
ISBN 978-3-11-128245-9, e-ISBN (PDF) 978-3-11-128434-7,
e-ISBN (EPUB) 978-3-11-128449-1

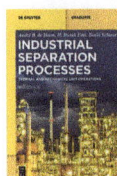

Industrial Separation Processes.
Thermal Unit Operations and Mechanical Unit Operations
André B. de Haan, H. Burak Eral, Boelo Schuur, 2025
ISBN 978-3-11-106363-8, e-ISBN (PDF) 978-3-11-106381-2,
e-ISBN (EPUB) 978-3-11-106408-6

Guillem Gilabert-Oriol

Biofouling and Organic Fouling

in Reverse Osmosis Membrane Elements

DE GRUYTER

Author
Dr. Guillem Gilabert-Oriol
Universitat Rovira i Virgili (URV)
and
DuPont Water Solutions
Tarragona, Catalonia
Spain
guillem.gilabertoriol@dupont.com
https://www.linkedin.com/in/guillemgilabert/

ISBN 978-3-11-163736-5
e-ISBN (PDF) 978-3-11-163907-9
e-ISBN (EPUB) 978-3-11-164033-4

Library of Congress Control Number: 2025931499

Bibliographic information published by the Deutsche Nationalbibliothek.
The Deutsche Nationalbibliothek lists this publication in the Deutsche Nationalbibliografie;
detailed bibliographic data are available on the internet at http://dnb.dnb.de.

© 2025 Walter de Gruyter GmbH, Berlin/Boston, Genthiner Straße 13, 10785 Berlin
Cover image: wacomka/iStock/Getty Images Plus.
Typesetting: Integra Software Services Pvt. Ltd.

www.degruyter.com
Questions about General Product Safety Regulation:
productsafety@degruyterbrill.com

Acknowledgments

To my colleague Gerard Massons, for having studied biofouling together for over a decade.

https://doi.org/10.1515/9783111639079-202

Preface

I still remember the day vividly. It was 2019, and I was in Dubai, visiting a customer at their reverse osmosis seawater desalination plant in the United Arab Emirates. They were grappling with serious problems caused by biofouling, an issue that had become a persistent challenge in their operations. As I stood there, interpreting the plots of normalized permeate flow, normalized salt rejection, and pressure drop they shared with me, the problem and its causes were crystal clear in my mind. Since 2013, I had dedicated myself to studying biofouling and organic fouling in reverse osmosis membranes. This focus allowed me to recognize patterns and solutions with ease. Eager to help, I wanted to share everything I knew with the customer so they could improve their plant's operations and experience peace of mind, free from the endless hassles fouling imposes on operators. But in that moment, I realized something important: there was no single, comprehensive resource I could point to, no book or guide that captured all the plots, equations, and insights I had accumulated over years of study. That realization sparked an idea: to write a book that would bring together all this knowledge in one accessible place. A resource not just for plant operators but also for the broader scientific community. This book is the result of that idea. It is structured into sixteen chapters, each designed to stand on its own for quick consultation, yet together they tell a cohesive story. The journey begins with the basics and fundamentals, laying the groundwork by explaining the key equations that govern fouling. From there, it progresses to explore the critical parameters that influence biofouling and the innovative technologies developed to address it, following their evolution through the stages of technological readiness. By understanding these parameters, we can devise effective solutions to mitigate and even prevent biofouling in reverse osmosis membranes. My hope is that this book serves as a valuable tool for anyone facing these challenges, helping them navigate the complexities of fouling with greater confidence and clarity.

Dr. Guillem Gilabert-Oriol
January 11, 2025

https://doi.org/10.1515/9783111639079-203

Abstract

Since their introduction in the 1970s, reverse osmosis and nanofiltration membranes have revolutionized desalination and water treatment. However, biological fouling, or biofouling, remains a persistent and critical challenge. Biofouling occurs when bacteria adhere to membrane surfaces and form biofilms. If left unchecked, these rapidly growing biofilms can compromise the mechanical integrity of membrane elements, causing irreversible damage. Additionally, organic fouling, caused by the deposition of organic matter on membranes, often intertwines with biofouling, further reducing membrane performance and permeability. This book explores the mechanisms behind biological fouling, the factors that influence its development, and practical strategies for its mitigation and prevention. By shedding light on these interconnected phenomena, it provides valuable insights for ensuring the efficient and reliable operation of reverse osmosis systems in desalination and water treatment plants.

https://doi.org/10.1515/9783111639079-204

Contents

Chapter 15
Limiting nutrients to prevent biofouling in seawater: part 1 —— 209

Chapter 16
Limiting nutrients to prevent biofouling in seawater: part 2 —— 225

About the author

Dr. Guillem Gilabert-Oriol holds a Master's Degree in Chemical Engineering and Processes from Rovira i Virgili University, earned in 2008. He began his career as a Research and Development Scientist at The Dow Chemical Company in 2010, working at the Global Water Technology Center in Tarragona, Catalonia, Spain. He completed his PhD in Ultrafiltration and took on leadership of the Reverse Osmosis Antifouling team in 2013.

Currently, Dr. Gilabert-Oriol serves as the Global Desalination Research & Development Leader at DuPont Water Solutions, a position he has held since 2019. In 2023, he became the head of the Tarragona Research & Development team. In addition to his role at DuPont, he is an Assistant Professor at Rovira i Virgili University in Tarragona, where he shares his expertise in water treatment technologies.

A passionate advocate for water management, Dr. Gilabert-Oriol is dedicated to the crucial role water plays in sustaining human life, improving quality of life, and fostering sustainable economic development.

https://doi.org/10.1515/9783111639079-206

Abbreviations

AFM	atomic-force microscopy
ANOVA	analysis of the variance
AOC	assimilable organic carbon
AS	antiscalant
ATP	adenosine triphosphate
BCA	bicinchoninic acid method
BEOP	biofilm enhanced osmotic pressure
BOD5	biological oxygen demand
BSA	bovine serum albumin
BV/h	bed volumes per hour
BW	brackish water
BW	backwash
CA	cellulose acetate
CAPEX	capital expenses
CDC	Centers for Disease Control and Prevention
CDOC	hydrophilic dissolved organic carbon
CF	cartridge filter
CH	carbohydrates
CIP	cleaning in place
CLSM	confocal laser scanning microscopy
COD	chemical oxygen demand
CTC	5-cyano-2,3-ditolyl tetrazolium chloride
cTEP	colloidal transparent exopolymer particles
CV	biofilm surface coverage
DBNPA	2,2-dibromo-3-nitrilopropionamide
DOC	dissolved organic carbon
dP	pressure drop
EPS	extracellular polymeric substances
F	flux
GFD	US gallons/(square feet · day)
GPD	US gallons/day
GPH	US gallons/hour
GPM	US gallons/minute
GFRP	glass fiber-reinforced-polymer
GWTC	Global Water Technology Center
HOC	hydrophobic dissolved organic carbon
L	liters
LC-OCD-OND	liquid chromatography – organic carbon detection – organic nitrogen detection
LDP	low pressure drop
LMH	liters / (square meter · hour)
LMW	low molecular weight
LOI	loss on ignition
LOQ	limit of quantification
MBR	membrane bioreactor
MF	microfiltration
mil	0.001 inch
MRI	magnetic resonance imaging

https://doi.org/10.1515/9783111639079-207

N	nitrogen
NF	nanofiltration
NOM	natural organic matter
OCT	optical coherence tomography
OPEX	operating expenses
P	pressure
P	phosphorous
Π	osmotic pressure
PN	proteins
ppb	µg/L
ppm	mg/L
Psi	pounds-force/inch^2
PV	pressure vessel
R	recovery
Q	flow
r^2	coefficient of determination
RO	reverse osmosis
SEWA	Sharjah Electricity and Water Authority
SMBS	sodium metabisulfite
SP	salt passage
SR	salt rejection
STD	standard
T	temperature
TCF	temperature correction factor
TEP	transparent exopolymer particles
TFC	thin-film composite
TDS	total dissolved solids
TIN	total inorganic nitrogen
TN	total nitrogen
TOC	total organic carbon
TON	total organic nitrogen
TSS	total suspended solids
UF	ultrafiltration
WW	wastewater
WWTP	wastewater treatment plant

Aim of this book

Book hypothesis

The hypothesis of this book is that biofouling and organic fouling can be severely mitigated by understanding which key factors affect their development.

Key hypothesis

Chapter 1 – Reverse osmosis fundamentals

Reverse osmosis is the most effective technology to desalinate water.
Data normalization is a key tool to understand if a reverse osmosis membrane system operates as intended or if it is suffering from fouling.

Chapter 2 – Biofouling fundamentals

Biofouling is characterized by an initial phase of a flat pressure drop increase over time, followed by a second phase consisting of exponential growth.
Organic fouling is characterized by a sudden loss in normalized permeate flow that later tends to stabilize and plateau. This phase is typically characterized by a flat pressure drop if no biofouling occurs at the same time.

Chapter 3 – Membrane fouling simulators: a quick tool to study biofouling and why biofouling does not depend on flux

The membrane fouling simulator can be used to study biofouling with a small investment.
Membrane fouling simulators are able to mimic biofouling happening in real installations.
Biofouling does not depend on water flux.

Chapter 4 – The importance of the feed spacer to prevent biofouling

Feed spacers play a key role in preventing biofouling.
Without a feed spacer, biofouling hardly grows.
Fouling-resistant membrane chemistries can mitigate biofouling.

https://doi.org/10.1515/9783111639079-208

Chapter 5 – The effect of the temperature and development of quick biofouling test

There is a temperature threshold that is needed for biofouling to start growing effectively.
Biofouling development depends on the availability of bioassimilable nutrients.
Biofouling can be accelerated by dosing bioassimilable nutrients.

Chapter 6 – Differentiating between biofouling and organic fouling during operation

The first fouling period is typically an organic fouling period.
The second fouling period is typically a biological fouling period.
Fouling is typically a combination of multiple fouling types.

Chapter 7 – Differentiating between biofouling and organic fouling on a membrane surface

A biofilm is mainly composed of EPS.
Biofouling can be distinguished from organic fouling by examining the portion of EPS in the entire fouling.
Biofouling development depends on the availability of bioassimilable nutrients.

Chapter 8 – Biofouling starts to develop before pressure drop starts to increase

TOC is quickly deposited on the membrane surface, leading to organic fouling.
ATP starts to accumulate after TOC has already accumulated.
ATP starts to accumulate on the membrane surface before the pressure drop begins to increase, ultimately leading to biofouling.
ATP on a membrane surface might be used as an early indicator to predict biofouling before it starts to develop.

Chapter 9 – Biofouling mainly happens in the lead elements

Biofouling mainly occurs in the lead elements.
It is very difficult to clean to the initial pressure drop once biofouling develops.
It is very difficult to fully remove a biofilm inside a reverse osmosis membrane.

Biocide cleaning can reduce the number of bacteria in a biofilm but does not help in preventing biofouling.

Biofouling can be easily brushed off or detached from the membrane and the feed spacer.

Organic fouling develops before biofouling starts.

Smart arrangement of feed spacers in a pressure vessel can reduce biofouling.

Chapter 10 – Biocides do not fully prevent biofouling

Dosing biocide allows the study of the initial organic fouling phase without biofouling interference.

Using biocides does not fully prevent biofouling.

Shock-dosing biocide does not fully prevent biofouling.

Fouling-resistant membrane chemistries can mitigate biofouling.

Chapter 11 – Visualizing biofouling on the membrane surface

Biofouling can be easily visualized and quantified on a membrane.

Biofouling mainly grows on the front elements in a pressure vessel.

Chapter 12 – Why preventing biofouling matters

A fouling-resistant element can reduce the number of chemical cleanings per year.

A fouling-resistant element can extend the operating time with the same number of cleanings.

Chapter 13 – Limiting nutrients to prevent biofouling in brackish water

Biofouling in brackish water can be effectively limited by restricting the bioassimilable nutrients that reach the reverse osmosis system.

The same biofouling prevention technology can reduce total suspended solids.

This biofouling prevention system uses biomass, not a biofilm, to prevent biofouling.

Chapter 14 – Limiting nutrients to prevent biofouling in wastewater

Biofouling in wastewater can be effectively limited by restricting the bioassimilable nutrients that reach the reverse osmosis system.

The same biofouling prevention technology can reduce cartridge filter replacements.
The same biofouling prevention technology can reduce organic compounds.

Chapter 15 – Limiting nutrients to prevent biofouling in seawater: part 1

Biofouling in seawater can be effectively limited by restricting the bioassimilable nutrients that reach the reverse osmosis system.
The same biofouling prevention technology can reduce cartridge filter replacements.

Chapter 16 – Limiting nutrients to prevent biofouling in seawater: part 2

Biofouling in seawater can be effectively limited by restricting the bioassimilable nutrients that reach the reverse osmosis system.

Key conclusions with key figures

Chapter 1 – Reverse osmosis fundamentals

Reverse osmosis is the most effective technology to desalinate water.

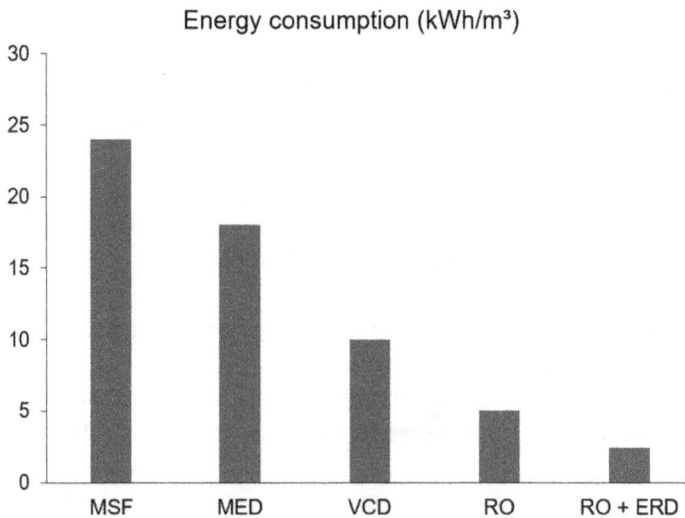

Figure 1: Energy consumption of different desalination technologies.

Data normalization is a key tool to understand if a reverse osmosis membrane system operates as intended or if it is suffering from fouling.

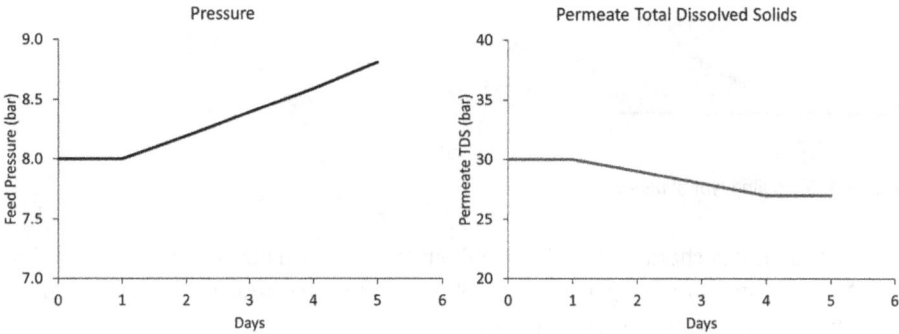

Figure 2: Feed pressure and permeate total dissolved solids' evolution.

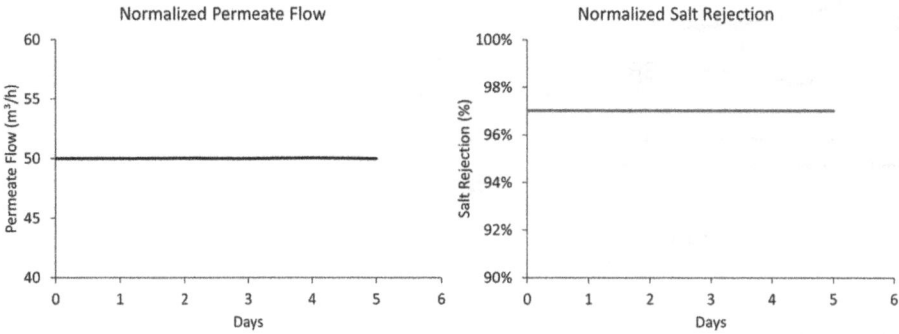

Figure 3: Normalized permeate flow and normalized salt rejection evolution.

Chapter 2 – Biofouling fundamentals

Biofouling is characterized by an initial phase of a flat pressure drop increase over time, followed by a second phase consisting of exponential growth.

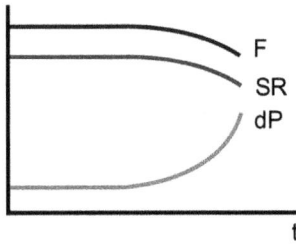

Figure 4: Biofouling typical behavior.

Organic fouling is characterized by a sudden loss in normalized permeate flow that later tends to stabilize and plateau. This phase is typically characterized by a flat pressure drop if no biofouling occurs at the same time.

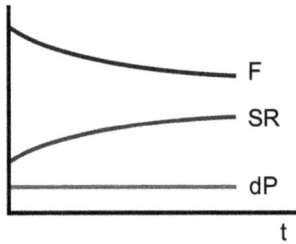

Figure 5: Organic fouling typical behavior.

Chapter 3 – Membrane fouling simulators: a quick tool to study biofouling and why biofouling does not depend on flux

The Membrane Fouling Simulator can be used to study biofouling with a small investment.

Membrane fouling simulators are able to mimic biofouling happening in real installations.

Biofouling does not depend on water flux.

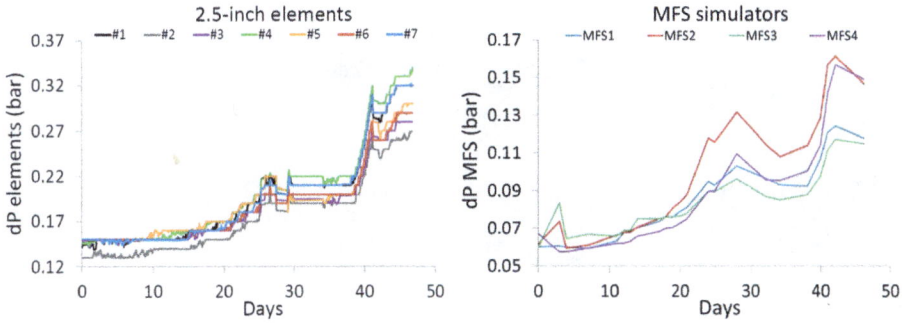

Figure 6: Pressure drop of MFS and 2.5-inch elements.

Chapter 4 – The importance of the feed spacer to prevent biofouling

The feed spacer plays a key role in preventing biofouling.
Without a feed spacer, biofouling hardly grows.

MFS cell with 34 mil spacer   **MFS cell without spacer**

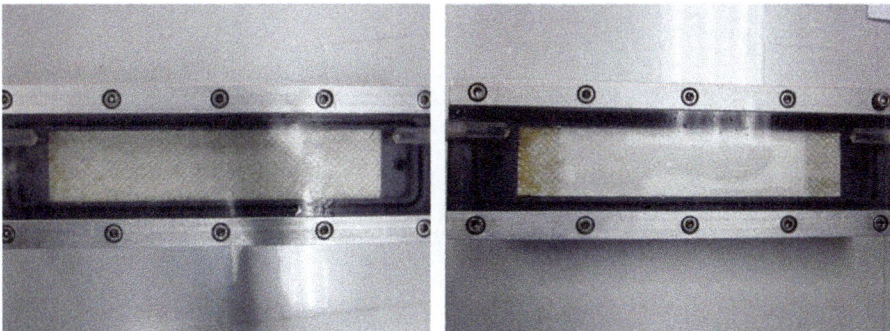

Figure 7: Biofilm distribution with and without a 34 mil spacer in the MFS cells.

Fouling-resistant membrane chemistries can mitigate biofouling.

| Membrane A (28 mil) | Membrane B (28 mil) | Membrane A (No spacer) |

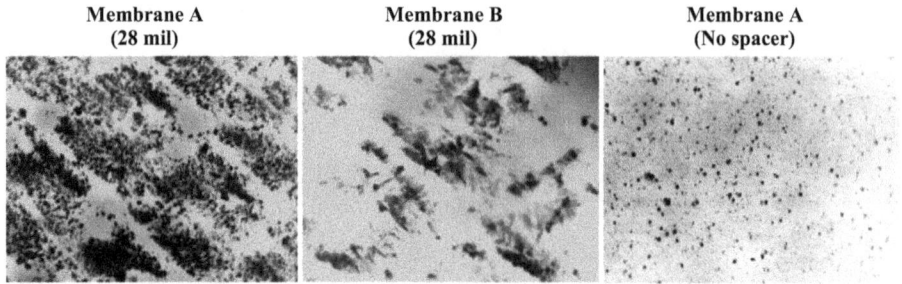

Figure 8: Biofouling developed on the membrane surface of Flat Cells.

Chapter 5 – The effect of the temperature and development of quick biofouling test

There is a temperature threshold that is needed for biofouling to start growing effectively.

Biofouling development depends on the availability of bioassimilable nutrients.

Biofouling can be accelerated by dosing bioassimilable nutrients.

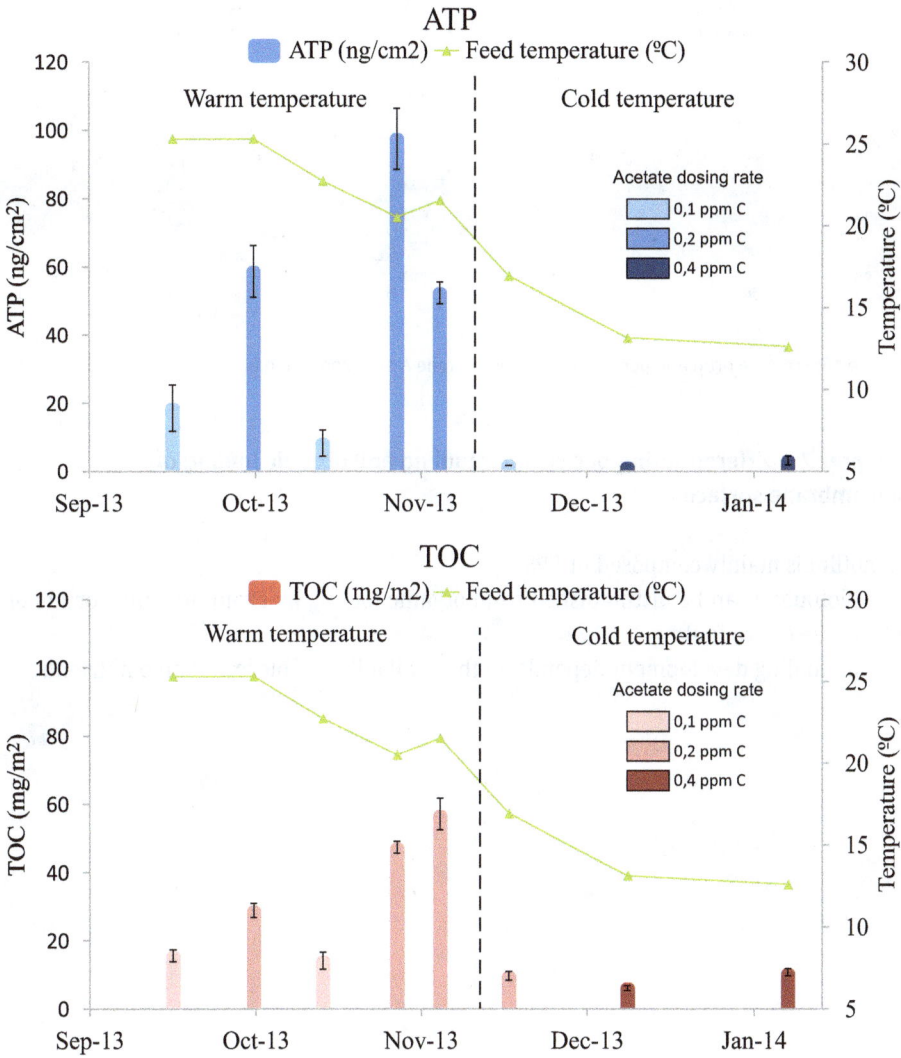

Figure 9: ATP and TOC evolution considering seasonal effect.

Chapter 6 – Differentiating between biofouling and organic fouling during operation

The first fouling period is typically an organic fouling period.
    The second fouling period is typically a biological fouling period.
    Fouling is typically a combination of multiple fouling types.

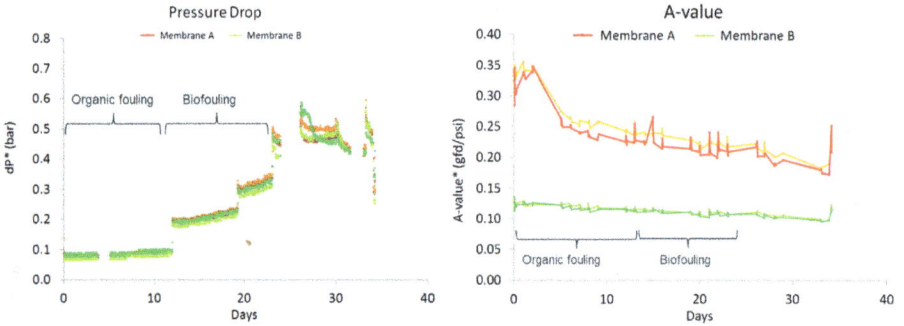

Figure 10: Pressure drop and permeability for Membrane A and Membrane B.

Chapter 7 – Differentiating between biofouling and organic fouling on a membrane surface

A biofilm is mainly composed of EPS.

Biofouling can be distinguished from organic fouling by examining the portion of EPS in the entire fouling.

Biofouling development depends on the availability of bioassimilable nutrients.

EPS (in TOC)

Pressure Drop

EPS (in TN)

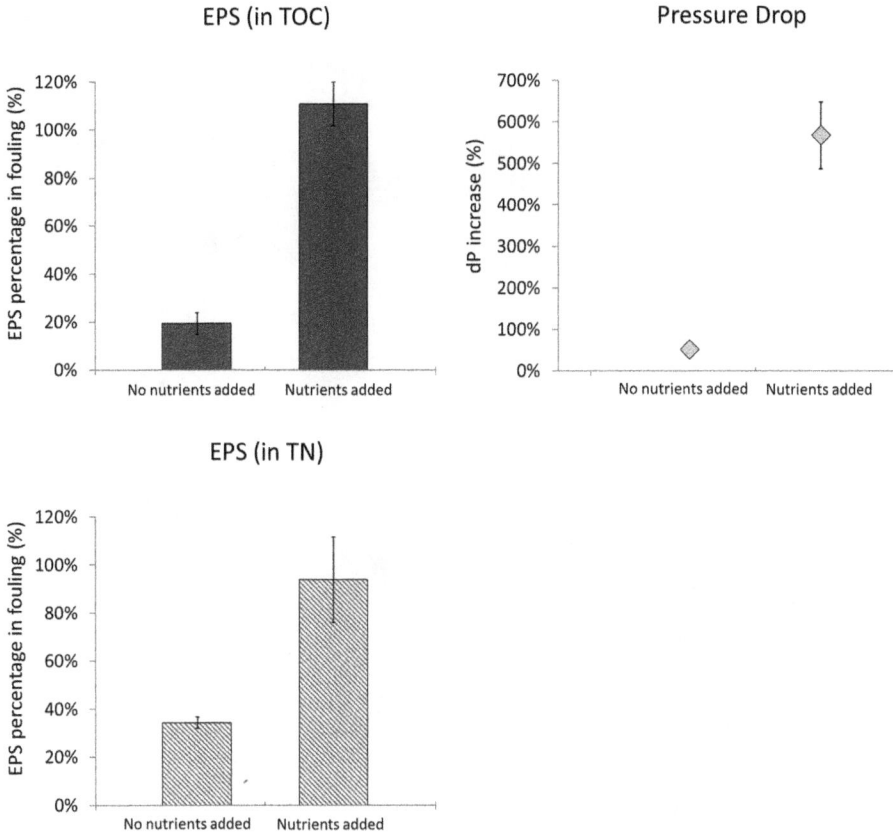

Figure 11: Correlation of membrane performance and EPS fraction for the two testing conditions.

Chapter 8 – Biofouling starts to develop before pressure drop starts to increase

TOC is quickly deposited on the membrane surface, leading to organic fouling.

ATP starts to accumulate after TOC has already accumulated.

ATP starts to accumulate on the membrane surface before the pressure drop begins to increase, ultimately leading to biofouling.

ATP on a membrane surface might be used as an early indicator to predict biofouling before it starts to develop.

ATP

$y = -1.8004x^2 + 10.924x - 0.5256$
$R^2 = 0.96$

ATP (ng/cm²)

Days

Pressure Drop

$y = 0.0119x^2 - 0.0029x + 0.0395$
$R^2 = 0.9832$

dP* (bar)

Days

TOC

$y = -5.1557x^2 + 33.006x$
$R^2 = 0.8762$

ATP (mg/m²)

Days

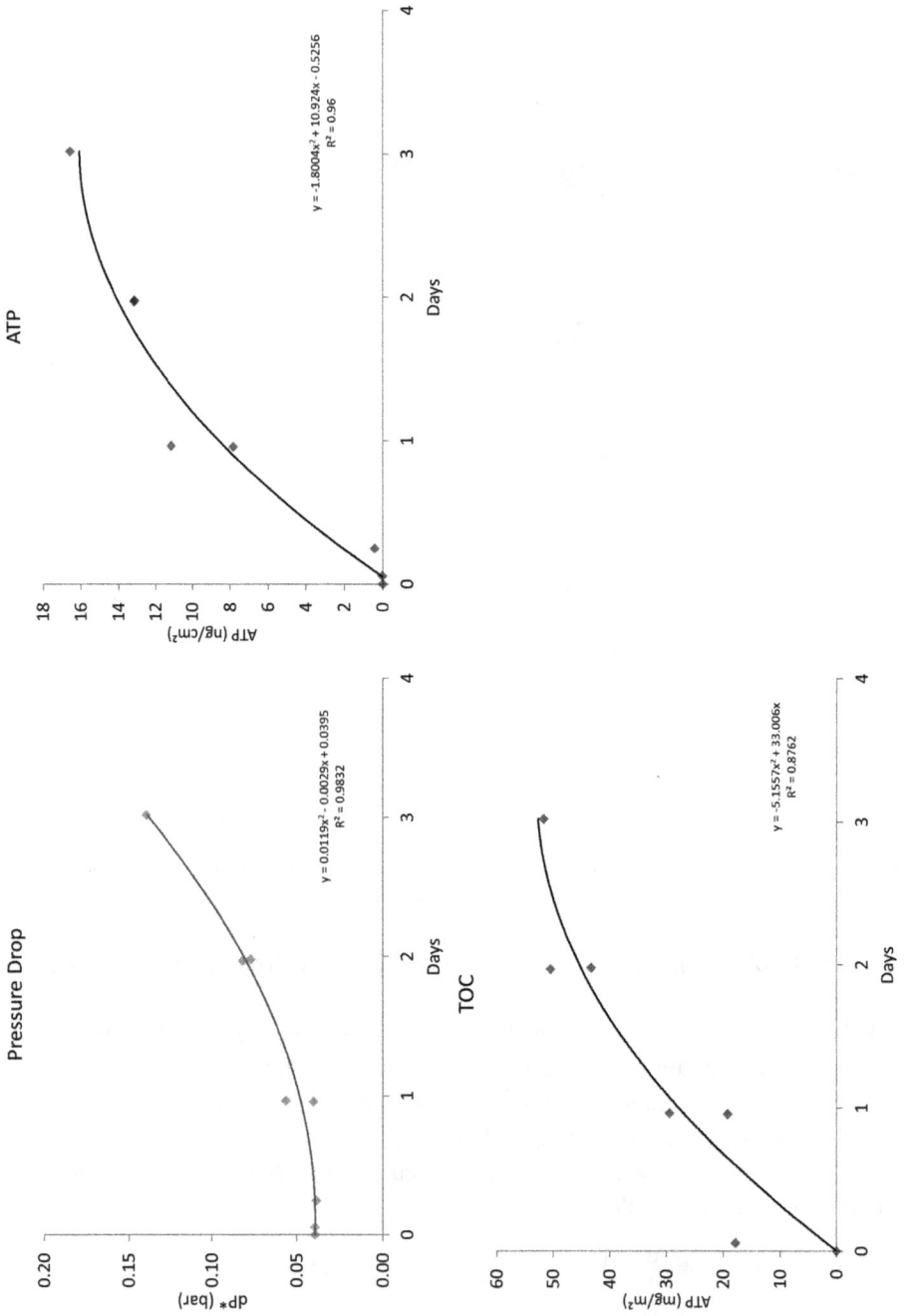

Figure 12: Pressure drop, ATC, and TOC of each membrane analyzed.

Chapter 9 – Biofouling mainly happens in the lead elements

It is very difficult to clean to the initial pressure drop once biofouling develops.
　　It is very difficult to fully remove a biofilm inside a reverse osmosis membrane.
　　Smart arrangement of feed spacers in a pressure vessel can reduce biofouling.

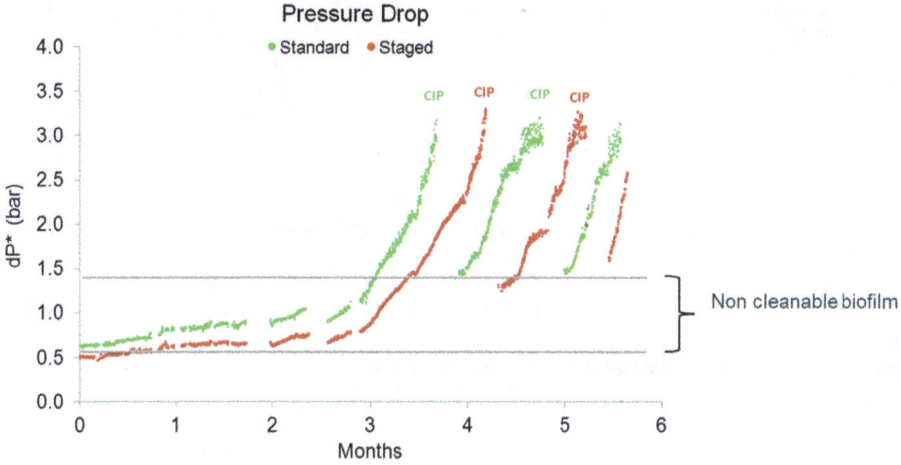

Figure 13: Pressure drop evolution.

Organic fouling develops before biofouling starts.

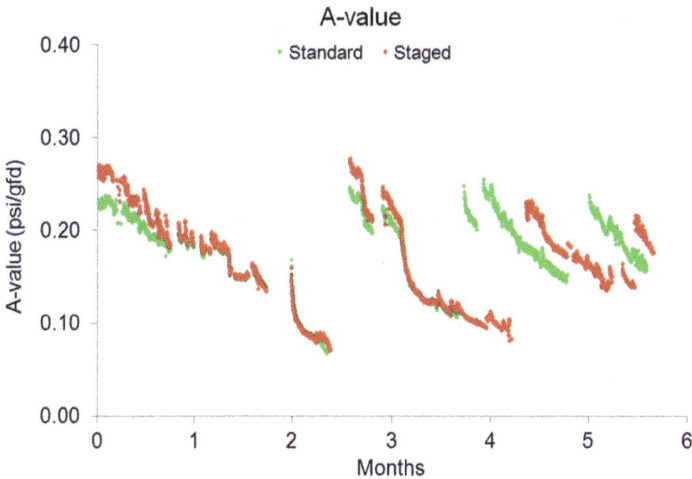

Figure 14: Water permeability (A-value) evolution.

Biofouling mainly occurs in the lead elements.

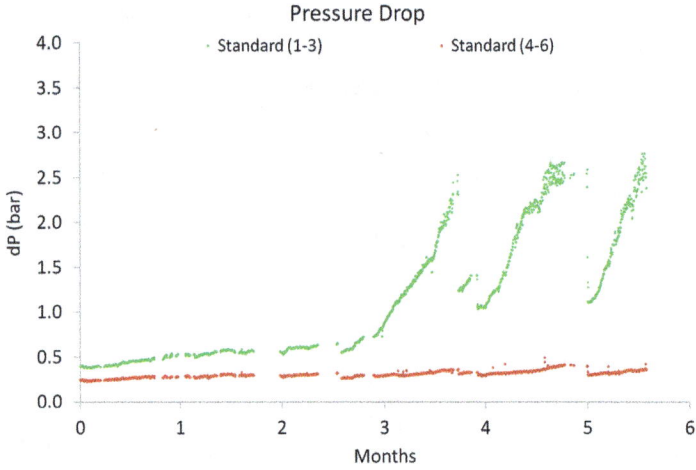

Figure 15: Pressure drop evolution in the tail and lead elements.

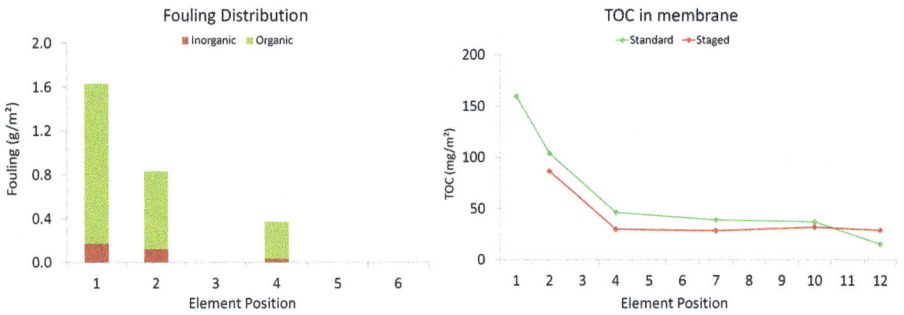

Figure 16: Fouling distribution across the pressure vessel.

Biocide cleaning can reduce the number of bacteria in a biofilm but does not help in preventing biofouling.

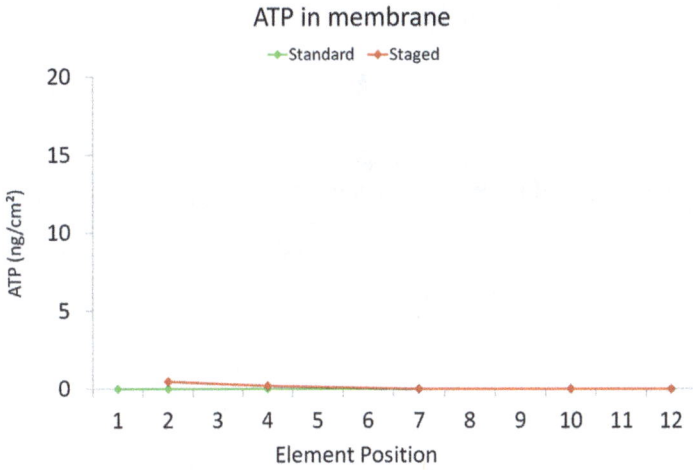

Figure 17: ATP distribution.

Biofouling can be easily brushed or detached from the membrane and the feed spacer.

Figure 18: Membrane and feed spacer before and after the max flushing test, and after being manually brushed.

Figure 19: Scraping the biofilm.

Chapter 10 – Biocides do not fully prevent biofouling

Dosing biocide allows the study of the initial organic fouling phase without biofouling interference.

Using biocides does not fully prevent biofouling.

Shock-dosing biocide does not fully prevent biofouling.

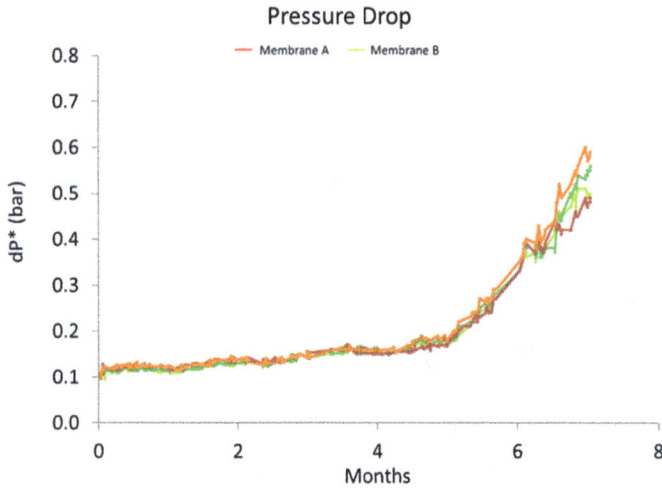

Figure 20: Normalized pressure drop evolution for Membrane A and Membrane B elements.

Fouling-resistant membrane chemistries can mitigate biofouling.

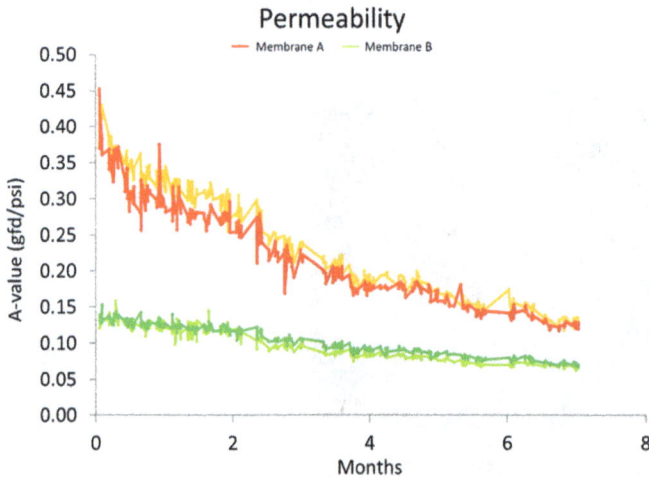

Figure 21: Normalized A-value evolution for Membrane A and Membrane B elements.

Chapter 11 – Visualizing biofouling on the membrane surface

Biofouling can be easily visualized and quantified on a membrane.

Biofouling mainly grows on the front elements in a pressure vessel.

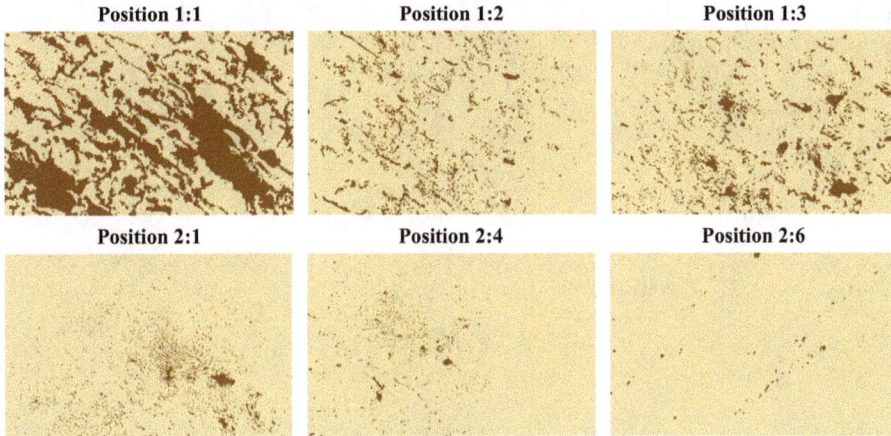

Figure 22: Biofouling distribution in a two-stage system.

Chapter 12 – Why preventing biofouling matters

A fouling-resistant element can reduce the number of chemical cleanings per year.

A fouling-resistant element can extend the operating time with the same number of cleanings.

Pressure Drop

• FilmTec™ XLE-440

Pressure Drop

• Innovation

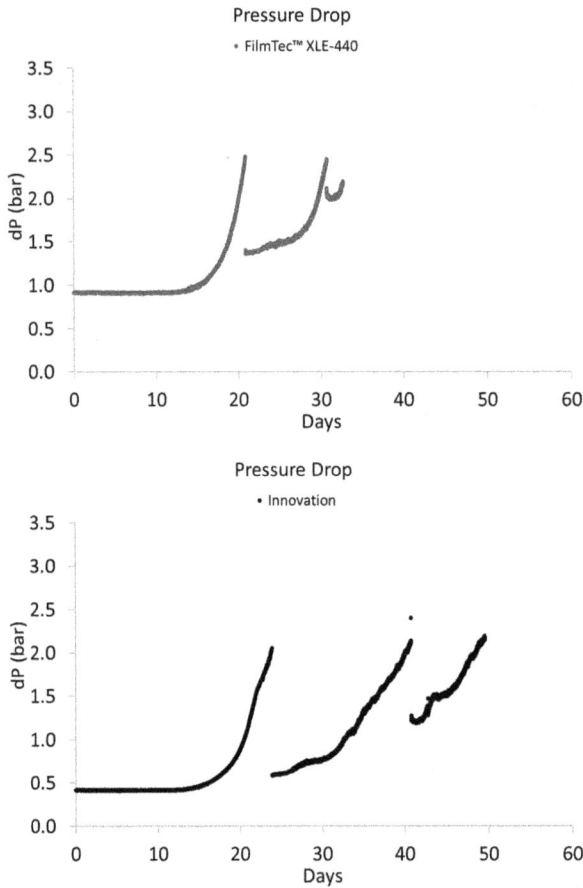

Figure 23: More operating time achieved thanks to a fouling-resistant membrane.

Chapter 13 – Limiting nutrients to prevent biofouling in brackish water

Biofouling in brackish water can be effectively limited by restricting the bioassimilable nutrients that reach the reverse osmosis system.

The same biofouling prevention technology can reduce total suspended solids.

This biofouling prevention system uses biomass, not a biofilm, to prevent biofouling.

Pressure drop in RO without DuPont™ B-Free™

Pressure drop in RO with DuPont™ B-Free™

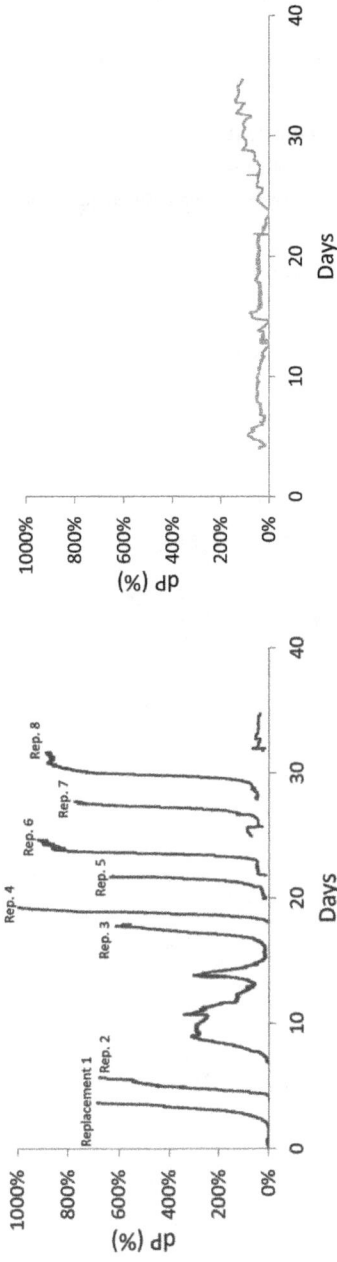

Figure 24: Reverse osmosis pressure drop without (left) and with (right) DuPont™ B-Free™.

Chapter 14 – Limiting nutrients to prevent biofouling in wastewater

Biofouling in wastewater can be effectively limited by restricting the bioassimilable nutrients that reach the reverse osmosis system.

The same biofouling prevention technology can reduce cartridge filter replacements.

The same biofouling prevention technology can reduce organic compounds.

Pressure drop in RO without DuPont™ B-Free™

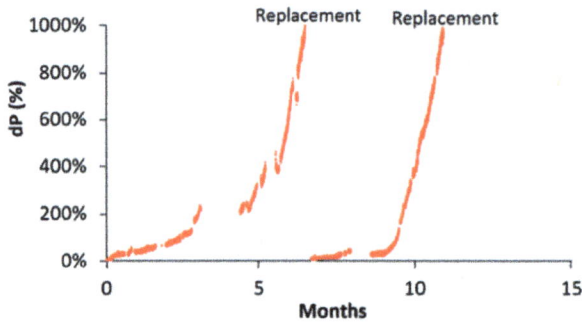

Pressure drop in RO with DuPont™ B-Free™

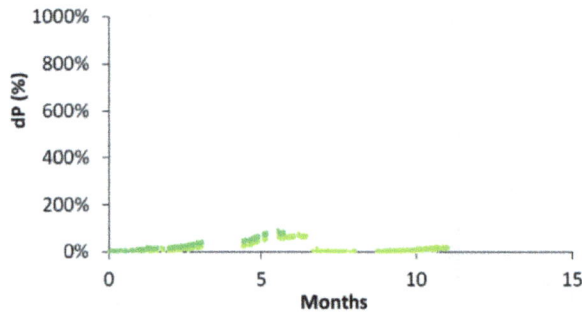

Figure 25: Reverse osmosis pressure drop without (top) and with (bottom) DuPont™ B-Free™.

Chapter 15 – Limiting nutrients to prevent biofouling in seawater: part 1

Biofouling in seawater can be effectively limited by restricting the bioassimilable nutrients that reach the reverse osmosis system.

The same biofouling prevention technology can reduce cartridge filter replacements.

Pressure drop in RO without DuPont™ B-Free™

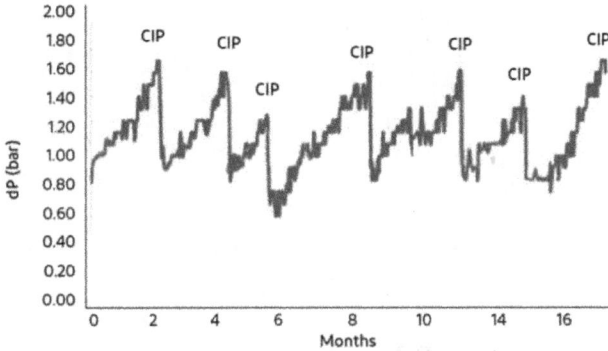

Pressure drop in RO unit with DuPont™ B-Free™

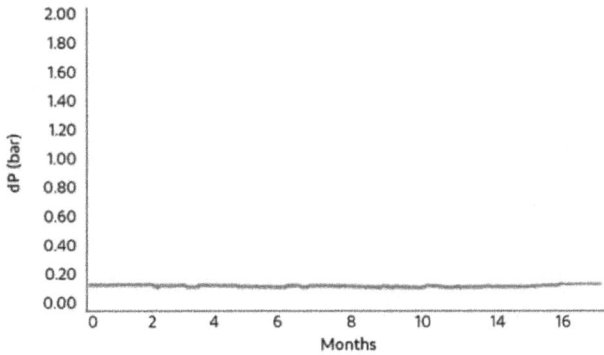

Figure 26: RO pressure drop evolution in seawater test.

Chapter 16 – Limiting nutrients to prevent biofouling in seawater: part 2

Biofouling in seawater can be effectively limited by restricting the bioassimilable nutrients that reach the reverse osmosis system.

Pressure drop in RO without DuPont™ B-Free™

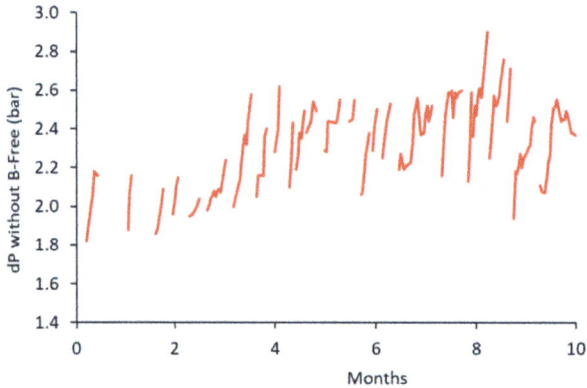

Pressure drop in RO with DuPont™ B-Free™

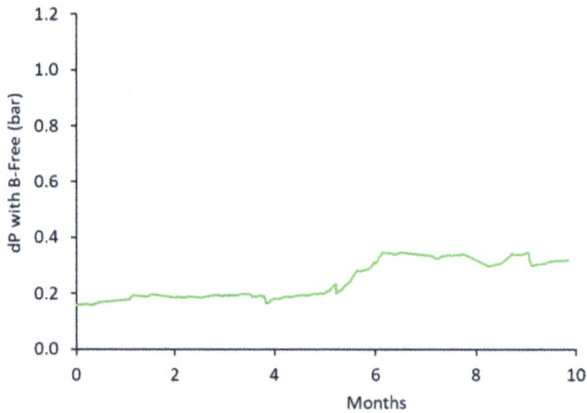

Figure 27: Cartridge filter pressure drop evolution in seawater test.

Chapters summary

Chapter 1 – Reverse osmosis fundamentals

This chapter explains the fundamental equations of reverse osmosis membranes. It details why its adoption has increased over the years and what its main advantages are compared to thermal processes. It explains what a reverse osmosis membrane is and what it is made from. It provides an overview of the main equations to calculate the osmotic pressure, the water flux, and the salt passing across a reverse osmosis membrane, and the main equations used to evaluate a reverse osmosis system. It also explains how to normalize plant operating data and why normalization matters to properly assess an installation's performance. These equations are the normalized permeate flow, the normalized salt rejection, and the normalized pressure drop. The main types of fouling found in a reverse osmosis membrane are detailed. These are biological fouling, also known as biofouling, organic fouling, inorganic fouling, also known as scaling, and particulate fouling.

Chapter 2 – Biofouling fundamentals

This chapter explains the fundamentals of organic and biological fouling. Biofouling is characterized by an initial phase of a flat pressure drop increase over time, followed by a second phase consisting of exponential growth. Organic fouling is characterized by a sudden loss in normalized permeate flow that later tends to stabilize and plateau. This phase is typically characterized by a flat pressure drop if no biofouling occurs at the same time. Special emphasis is placed on understanding which factors most influence the development of biofouling, along with explaining the strategies that can be taken to prevent it. This is illustrated by the concept of the biofouling triangle, where three key factors in biofouling development are identified. These factors are temperature, which promotes bacterial growth; bacteria availability, which is needed for bacteria to reproduce; and nutrient availability, which is needed for bacteria to grow and reproduce, and comprises assimilable organic nutrients such as carbon, nitrogen, and phosphorus, together with oxygen, which is used for bacteria to grow.

Chapter 3 – Membrane fouling simulators: a quick tool to study biofouling and why biofouling does not depend on flux

This chapter presents a research tool, the Membrane Fouling Simulators. These small assets are a quick way to study biofouling. In this chapter, it is explained how these tools are representative of the biofouling that occurs in larger installations. This is demonstrated through validation with two different plants. This tool is used to study

new feed spacers with the potential to reduce the impact of biofouling in reverse osmosis membranes. This chapter also outlines that permeate flux does not appear to
be a relevant parameter for biofouling development, as Membrane Fouling Simulators operate under no permeate water flux across the membrane and are able to fully
mimic the biofouling development.

Chapter 4 – The importance of the feed spacer to prevent biofouling

This chapter outlines the importance of the feed spacer in preventing biofouling. This
is proven across multiple experiments, where it can be seen that without a feed
spacer, biofouling hardly develops. This chapter also outlines how different membranes with different properties can mitigate the biofouling development in a reverse
osmosis membrane element. It is hypothesized that the feed spacer plays a crucial
role in preventing biofouling, as biofouling needs a place to adhere and grow that is
sheltered from the water flow but still receives enough nutrients to sustain itself
and grow.

Chapter 5 – The effect of the temperature and development of quick biofouling test

This chapter outlines the importance of temperature in developing biofouling, where
it is shown that when the temperature dropped below 17 °C, biofouling developed at a
much slower pace. It also validates and establishes a quick biofouling test where dosing essential bioassimilable nutrients made of carbon, nitrogen, and phosphorus
speeds up the development of biofouling. Finally, this quick biofouling test was
adapted so that biofouling could properly be studied at lower temperatures. Dosing
nutrients can be used to accelerate the biofouling period and enables better study of
biological fouling.

Chapter 6 – Differentiating between biofouling and organic fouling during operation

This chapter compares two different membranes. Both membranes are compared
side by side during an initial organic fouling phase, followed by a second biological
fouling phase. In these two phases, it can be seen how the fouling-resistant membrane
allows for improved fouling resistance as it is able to withstand a lower water permeability loss over time.

Chapter 7 – Differentiating between biofouling and organic fouling on a membrane surface

This chapter outlines how biofouling can be distinguished from organic fouling when a membrane is analyzed together with the operating parameters. It can be seen that biofouling is identified by a significant increase in pressure drop. This is later confirmed during the membrane autopsy, when it is verified that the majority of the fouling present on the membrane is composed of exopolymeric substances (EPS). In contrast, organic fouling is identified by hardly increasing the pressure drop and having a smaller percentage of EPS in its total fouling amount when the membrane is analyzed. This is explained by the fact that when bacteria build biofilms, the biofilm, which is mainly composed of EPS, blocks the feed-concentrate channel, leading to a pressure drop increase. On the contrary, organic fouling is characterized by organic matter adsorbing to the membrane, leading to a sharp normalized flux decline (higher energy consumption), but not blocking the feed-concentrate channel and therefore not contributing to a pressure drop increase. Additionally, these organics might come from sources other than the biofilm, leading to a smaller percentage of EPS in the overall fouling composition.

Chapter 8 – Biofouling starts to develop before pressure drop starts to increase

This chapter shows how, despite that initially a biofilm might not be noticed by an increase in pressure drop over time, it might have already started forming. This is confirmed by the measurement of ATP accumulation on the membrane, where it is observed that ATP, which represents bacteria, starts increasing very fast at the beginning of the trial, when the pressure drop still remains quite flat. This research also shows how TOC starts accumulating very fast on the membrane at the beginning of the trial. This indicates the beginning of the first organic fouling period. This TOC accumulation on the membrane matches well with the mechanism by which biofilms are formed, as for a biofilm to start growing, it needs a substrate. This is referred to as the conditioning layer. This chapter also shows how ATP measurement on a membrane might be used as an early indicator of biofouling, pointing to a way to identify when a biofilm is starting to form, before this early biofilm formation leads to a biofilm that is big enough to start causing a decrease in pressure drop, therefore leading to the problem of biofouling in a reverse osmosis system. ATP starts to accumulate after TOC has already accumulated.

Chapter 9 – Biofouling mainly happens in the lead elements

This research shows how a smart distribution of feed spacers in a pressure vessel can reduce biofouling. It also shows that biofouling mainly develops in the lead elements. It also highlights the difficulty in cleaning to the initial pressure drop once biofouling develops. It shows that it is very difficult to fully remove a biofilm inside a reverse osmosis membrane. It also points out that caustic cleaning provides the highest cleaning effectiveness. It also shows that biocide cleaning can reduce the number of bacteria in a biofilm but does not help in preventing biofouling. It is also observed that biofouling can be easily brushed or removed by stirring in a beaker from a membrane and feed spacer. Finally, this trial shows the two classical fouling phases: the first organic fouling phase and the later biofouling phase that grows over the already organically fouled membrane.

Chapter 10 – Biocides do not fully prevent biofouling

This chapter shows how shock-dosing a non-oxidizing biocide initially helps delay biofouling, but after a certain amount of time, biofouling starts to develop normally. It can also be seen the role that a fouling-resistant membrane has in experiencing a smaller loss of water permeability over time. It is hypothesized that, despite the biocide being effective in eliminating bacteria, after the biofilm has grown enough, bacteria can shelter in the EPS, and this is when the biocide stops being effective. Dosing biocide is a useful method to study the first organic fouling part without biofouling interference.

Chapter 11 – Visualizing biofouling on the membrane surface

This chapter explains how biofouling can be easily visualized and quantified on a membrane surface. This technique is used to demonstrate that biofouling mainly grows on the first elements in a pressure vessel, where it can be seen which are the main areas that biofouling colonizes, as well as what percentage of the membrane is occupied by biofouling.

Chapter 12 – Why preventing biofouling matters

This chapter provides an example of why designing a fouling-resistant reverse osmosis membrane matters. In this particular case presented, it can be seen that thanks to the enhanced fouling-resistant properties of the newly developed element, chemical cleanings due to biofouling can be reduced by 30%, therefore extending the period the membranes can operate without cleanings by 56%. At the same time, energy consumption is reduced by 10%, while the feed-concentrate pressure drop is reduced by 55%.

Chapter 13 – Limiting nutrients to prevent biofouling in brackish water

This chapter explains how biofouling can be effectively prevented by adding a pretreatment step between the ultrafiltration and the reverse osmosis system. This pretreatment is called DuPont™ B-Free™ and focuses on limiting the nutrients that can reach the downstream reverse osmosis system, so that biofouling does not develop in the reverse osmosis system. Instead of biofouling growing in the lead elements of the reverse osmosis system, biofouling grows in this pretreatment, where it can be easily controlled and cleaned, and where its pressure drop performance can always be restored to its initial values. This chapter also showcases how this pretreatment technology can reduce up to 99.5% of suspended solids. Additionally, the relationship between biomass thickness and additional pressure drop increase is also established.

Chapter 14 – Limiting nutrients to prevent biofouling in wastewater

This chapter explores the effective prevention of biofouling by incorporating a pretreatment step between the ultrafiltration and reverse osmosis systems. This pretreatment stage is designed to minimize the nutrients available to the reverse osmosis system, thereby preventing biofouling from developing downstream. Unlike the biofouling growth typically observed in the lead elements of reverse osmosis systems, as discussed in previous chapters, biofouling is redirected to the pretreatment stage. Here, it can be more easily managed and cleaned, ensuring the system's pressure drop performance is consistently restored to its original levels. Additionally, the chapter demonstrates how this approach reduces the replacement frequency of cartridge filters. The technology behind this innovation, DuPont™ B-Free™, is shown to effectively prevent biofouling in wastewater environments while also reducing the risk of organic fouling in reverse osmosis systems.

Chapter 15 – Limiting nutrients to prevent biofouling in seawater: part 1

This chapter discusses a method to effectively prevent biofouling by incorporating a pretreatment step between the ultrafiltration and reverse osmosis systems. The purpose of this pretreatment is to limit the nutrients reaching the reverse osmosis system, thereby inhibiting the development of biofouling within it. Instead of biofouling occurring in the lead elements of the reverse osmosis system, as detailed in previous chapters, this process shifts biofouling to the pretreatment stage. In this controlled setting, biofouling can be easily managed, cleaned, and its pressure drop performance reliably restored to original levels. The chapter also highlights how this approach reduces the frequency of cartridge filter replacements. Using the example of a desalination plant in the Canary Islands, Spain, which is known for its severe biofouling issues, it demonstrates the effectiveness of DuPont™ B-Free™ technology in mitigating this challenge.

Chapter 16 – Limiting nutrients to prevent biofouling in seawater: part 2

This chapter provides another example of how the DuPont™ B-Free™ pretreatment technology is able to prevent biofouling by limiting the nutrients that reach the downstream reverse osmosis membranes in a seawater desalination plant in the United Arab Emirates.

Chapter 1
Reverse osmosis fundamentals

This chapter explains the fundamentals of reverse osmosis (RO) membranes. It details why its adoption has increased over the years, and what its main advantages are compared to thermal processes. It explains what a RO membrane is and what it is made from. It also gives an overview of the main equations that are used to characterize RO, and how they can be used. These are the equations to calculate the osmotic pressure, the equations needed to calculate the water flux across an RO membrane, and the salt passing through the membrane. This chapter also details the main equations used to evaluate an RO system. It is also explained how to normalize plant operating data and why normalization matters in order to properly assess an installation's performance. These equations allow calculating the normalized permeate flow, the normalized salt rejection, and the normalized pressure drop. Finally, the main types of fouling that can be found in an RO membrane are detailed. These are biological fouling, also known as biofouling, organic fouling, inorganic fouling, also known as scaling and particulate fouling.

1.1 Membrane filtration

Membranes are classified according to their pore diameter. An overview of each membrane technology regarding its pore diameter is given in the next paragraph. In addition, Figure 1.1 provides a graphical scheme summary [1]. Figure 1.2 details the intersection region between both mass transport models [1]. The pore flow model is represented by ultrafiltration (UF), and the solution diffusion model is represented by . In the intermediate section, nanofiltration (NF) combines both models to describe its behavior. Finally, Table 1.1 illustrates some examples of typical species that are filtrated using one of the described membrane technologies, together with their typical size [1]. Therefore, using Figures 1.1 and 1.2, it is possible to assess which filtration technology will be more suitable to filtrate or concentrate one of the species shown in the table.

RO membranes have pore diameters that range from 0.1 nm to 1 nm [1]. These pores have the particularity that they are so small that discrete pores do not exist. Instead, the pores are formed through unstable spaces between polymer chains, which are created and faded as a result of their molecular thermal motion. These fluctuating pores represent the diffusion of species throughout the dense membranes. In contraposition, the bigger and more stable pores observed in UF porous membranes represent the mass flux

Acknowledgments: A part of this chapter was originally published as a chapter with the following reference: G. Gilabert-Oriol, Reverse Osmosis and Nanofiltration, IWA Publishing (2024); from the book with the following reference: S. G. Salinas-Rodríguez, L.O. Villacorte, Experimental Methods for Membrane Applications in Desalination and Water Treatment, IWA Publishing (2024).

https://doi.org/10.1515/9783111639079-001

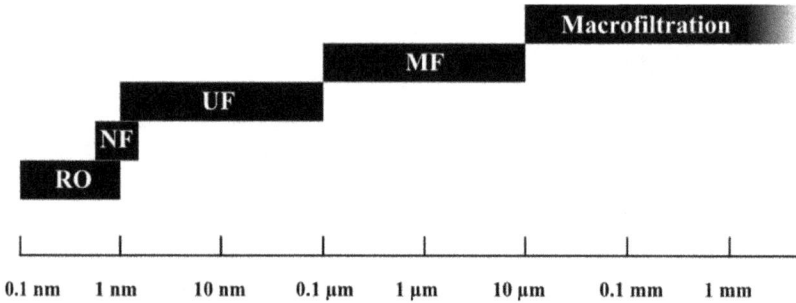

Figure 1.1: Membranes classified by their pore diameter.

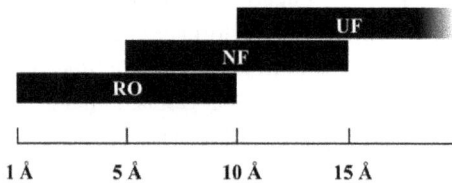

Figure 1.2: Reverse osmosis, nanofiltration, and ultrafiltration membranes classified by their pore diameter.

Table 1.1: Species with their size.

Species	Size
H_2O	0.2 nm
Na^+	0.37 nm
Sucrose	1 nm
Hemoglobin	7 nm
Influenza virus	0.1 μm
Pseudomonas diminuta	0.28 μm
Staphylococcus bacteria	1 μm
Starch	10 μm

through convection described by the pore flow model. The solution diffusion model, which is not covered in this book, makes two assumptions. The first is that the solvents dissolve inside the membrane, and thereafter they diffuse through the dense film according to the present concentration gradient. In RO, separation occurs because of the different solubility and mobility of each species throughout the membrane.

NF membranes have pore diameters that range from 0.5 nm to 1.5 nm [1]. These pores have the particularity of being between truly microporous membranes and clearly dense films. Therefore, mass transfer through NF membranes is described using both pore flow and solution diffusion models. This happens because if membrane polymer chains are very stiff, the molecular motion of the polymer is restricted, and semipermanent microcavities are formed, which are interconnected. These membranes are also called polymers with intrinsic microporosity [2].

UF membranes have pore diameters that range from 1 nm to 0.1 μm [3]. These pores have the property of being larger and more stable micropores, which do not vary over time and do not appear and disappear because of molecular thermal motion like RO membranes do. The filtration principle that produces separation is the sieving mechanism practiced by the pores, which results in a convective flux across the membrane.

Microfiltration (MF) membranes have pore diameters that range from 0.1 μm to 10 μm [4]. These pores are similar to the pores used by UF membranes but are much larger. Their filtration mechanism is also described by the pore flow model, as it achieves separation using the same sieving mechanism principle.

Macrofiltration, also known as conventional filtration, presents pore diameters above 10 μm [5]. The pore sizes are normally visible to the human eye, and they use the same sieving separation mechanism as UF and MF. Therefore, its flow across the membrane is also achieved by the pore flow model.

1.2 The rise of reverse osmosis

Water scarcity is being recognized as one of the main threats that mankind is facing globally [6]. RO membrane technology has developed as a promising technology to address this problem, holding roughly 44% market share and growing among all the desalinating technologies [7]. This increased market adoption has been driven as materials have been improved and costs have dropped [8]. This is especially relevant in arid regions such as in the Middle East countries (ME), where the population is located in arid and semi-arid regions, with very limited rainfall, and where, due to high ambient temperatures, evaporation contributes to a higher degree of stress on the naturally available water sources. Moreover, water scarcity is aggravated by the population increase this region is exposed to, as well as the economic development [9]. All these factors, together with the favorable energy-to-product quality ratio that seawater RO (SWRO) offers, have situated this technology as one key driver to sustain population living standards in ME countries [10].

The first technologies used to desalinate seawater employed thermal processes where seawater is evaporated, and then the steam, which is free of salts, is recondensed to obtain fresh water. These thermal-driven technologies that rely on distillation are multistage-flash distillation, multiple-effect distillation, and vapor compression desalination processes. The main drawback of these methods is the significant amount of energy required per cubic meter of water produced, compared to modern RO-based desalination. As Figure 1.3 shows, RO desalination, especially when coupled with energy recovery devices, is 10 times more energy-efficient than multistage flash desalination and 4 times more efficient than vapor compression distillation [11, 12].

Energy consumption (kWh/m³)

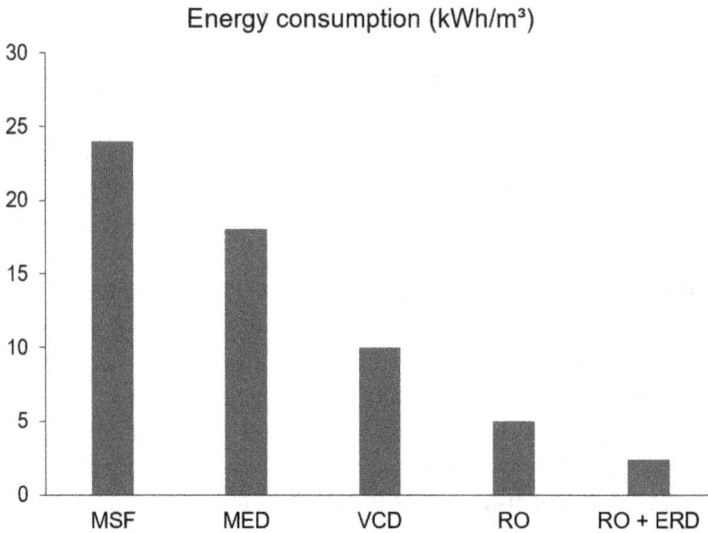

Figure 1.3: Energy consumption of different desalination technologies.

One of the key aspects that has allowed RO desalination to be so energy efficient is the introduction of energy recovery devices [13]. These systems are like heat exchanger operation units, but instead of exchanging thermal energy, they exchange pressure. This allows for the recovery of almost all the energy that was previously lost in the concentrate water stream of a SWRO system and uses it to pressurize the same volumetric flow in the feed of the RO system. If we take seawater desalination as an example, this can reduce the energy expense in a seawater desalination plant by 55%. Also, a smaller high-pressure pump is needed to pressurize the feed of an RO system, as 55% of the flow is already pressurized coming from the energy recovery device. It is worth mentioning that previously, the concentrate stream of an RO system that could come pressurized up to 80 bar was discharged into the open atmosphere, and all this energy was lost.

1.3 Sustainability of reverse osmosis

RO membrane technology offers a solution to achieve the sustainability development goals that the United Nations has set up for 2030. This has been stated and recognized by the United Nations [14]. Thanks to desalination technology, it is possible to fight against water scarcity and obtain water of high quality. If the energy is powered through renewable sources, such as solar power or photovoltaics, it is possible to achieve drinking water of high quality. This allows us to fight against climate change, as well as to provide an unlimited amount of drinking water for the population, which is not linked to whether it rains or not in nature.

It is also important to make sure that desalination concentrate discharge is done properly, and that this brine is properly managed through diffusors or by mixing it with seawater, so its discharge does not affect marine species or the environment [15].

The use of brine can also be a resource, to extract valuable minerals such as sodium chloride, magnesium compounds, bromide, and rubidium, among others. This is important as it allows for reducing the cost of water in desalination technologies, as well as preserving natural resources such as landscapes and mountains from invasive mining extraction operations [16].

Finally, some endeavors, such as the water positive initiative, aim to take the sustainability impact of desalination and water reuse one step further. This is inspired by the carbon credits system, but for water credits. It aims to help those companies that strive to become water neutral in terms of their water footprint, so that they can compensate for their water-negative use with those companies that are net producers of water, such as desalination and water reuse installations. This can help drive awareness of the importance of reducing the water footprint and help preserve this valuable resource, as well as make sustainable water treatment processes more affordable.

1.4 Understanding the osmosis process

In order to understand why RO has this name, it is important to first understand what osmosis means. Osmosis is a natural process that only takes place when there is a semipermeable membrane. A semipermeable membrane is defined by a solvent like water passing through it, and a solute like salt that cannot pass through the membrane.

1.4.1 Semipermeable membranes

A semipermeable membrane is defined by letting a solvent like water pass through it, but not letting a solute like salt pass through the membrane. In order to better understand this process, it can be useful to imagine a simplified scenario, where only water is considered as a solvent, and only sodium chloride (NaCl) is considered as a solute. When sodium chloride is dissolved in water, both sodium and chloride are separated in terms of Na^+ ions and Cl^- ions, as follows:

NaCl equilibrium,

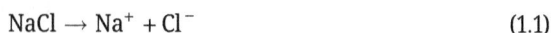

$$NaCl \rightarrow Na^+ + Cl^- \tag{1.1}$$

Although both the atomic radius of water molecules (H_2O) and Na^+ ions and Cl^- ions are similar, and therefore it could be expected that both water and sodium and chloride ions can pass through the membrane at a similar rate, this is not the case. The reason for this phenomenon is the solvation phenomenon. In order to maintain electrical neutrality, when ions are dissolved in water, they become surrounded by water molecules. Since water molecules are polar, they tend to orient their mostly negative charge with the sodium positively charged ions. The same happens for chloride negatively charged ions, as they get surrounded by the negatively charged side of the water molecules. The ultimate consequence for both chloride and sodium ions is that their effective sizes drastically increase as a result of being surrounded by water molecules. This is the main reason why small molecular weight species that are charged are much better rejected from a solute like water. Therefore, the smaller a species is, and the more charged it is, the better it will be rejected by an RO membrane.

The solvation effect is depicted in Figure 1.4, showing how water molecules (red for oxygen and gray for hydrogen) interact with sodium ions (blue) and chloride ions (green) as they pass through the RO membrane.

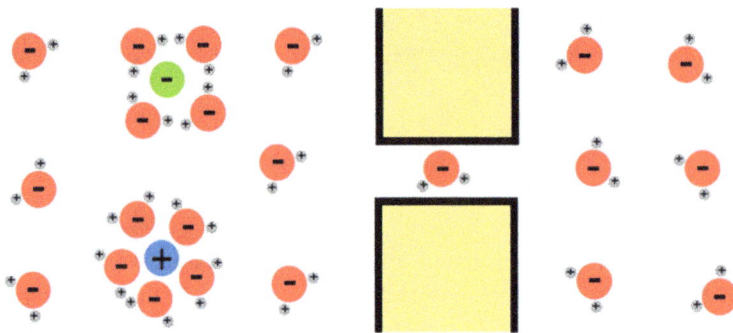

Figure 1.4: Solvation effect of water (red and gray molecules) on sodium ions (blue) and chloride ions (green) as they pass through a space in the reverse osmosis membrane.

It should be noted that this "pore" drawn in this diagram represents a space that is created in the polyamide chain. As polymers rotate as a result of being above 0° Kelvin temperature, small "pores" like the one drawn in the diagram are created, where water can pass through freely via a pressure-driven convective flow [17].

1.4.2 The reverse osmosis process

Once the concept of a semipermeable membrane is defined, and it remains clear that it lets a solvent like water pass, while a solute like sodium chloride cannot pass, the natural process of osmosis can be defined. As mentioned, the osmosis process only makes sense when a semipermeable membrane is present. Osmosis is defined as a natural process when a species that is at a higher concentration goes to dilute the other side of the membrane that has a lower concentration of this species. This is typically the case with solvents like water and solutes like salt. Water can move through a semipermeable membrane, while salt cannot. Therefore, water with lower salt concentration, like fresh water, will always go to dilute the water with higher salt concentration, like seawater, as this will have a lower water concentration. Since this natural phenomenon is not the goal of producing fresh drinking water, as it is not desired to get fresh water consumed, in order to achieve the opposite result and be able to generate fresh drinking water from seawater, the RO process was invented. This process consists of applying pressure on the seawater that is enough to reverse this process and being able to generate fresh water instead of consuming it.

Osmosis occurs in nature. One example is cherries. When it rains, rain droplets cover the cherry skin. The cherry skin acts like a semipermeable membrane; it lets the water pass through but not salts or sugars to escape. Therefore, after it rains, water travels from the outside of the skin, where the water concentration is higher, toward the inside of the cherry skin, where the water concentration is lower as a result of the fiber and fructose that cherries contain. The ultimate result is that as water from rain enters the inside of the cherry, it can eventually crack the cherry skin as its volume increases, and the cherry skin might not be able to expand properly before cracking to account for the increase in the cherries' volume.

In order to visualize this, consider two different solutions with the same volume each. These solutions can be put in contact through a semipermeable membrane. The first solution is seawater, which is assumed to have 30 g of salt with 970 g of water. This represents a 3.1% salt concentration. The second solution is fresh water, which is assumed to have 1 g of salt with 999 g of water. This represents a 0.1% salt concentration. This setup can be observed in Figure 1.5. In this case, the two solutions are separated. Fresh water will go to dilute seawater, while salt will not be able to pass.

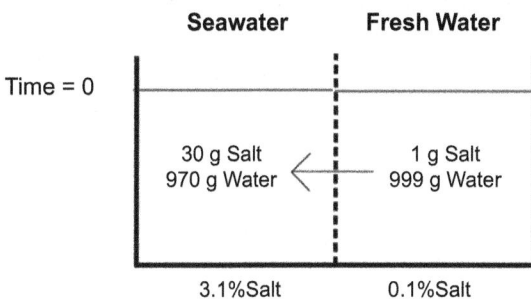

Figure 1.5: Initial experimental setup.

As salt cannot pass, water from the freshwater side will go to dilute the seawater. Water will keep passing until both sides' salt concentration is the same. This will happen after

935 g of water have passed from the freshwater side to the seawater side. At this equilibrium point, concentrations on both sides will be the same at 1.6%. It is important to notice that the salt mass on both sides is kept the same, but the water mass has changed. This change in water volume, which increases on the seawater side but decreases on the freshwater side, leads to a difference in water height between the seawater and fresh water side. This difference in height is what is called osmotic pressure. This osmosis process is shown in Figure 1.6.

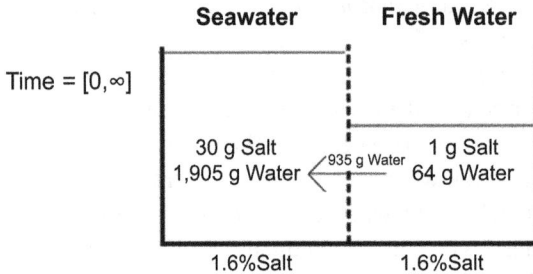

Figure 1.6: Osmosis process.

If one aims to reverse this naturally occurring osmosis process, one needs to apply a pressure that is at least the same as this osmotic pressure. Once this pressure is applied, the water flow is reversed. If exactly the same pressure as the osmotic pressure is applied, it will be possible to reverse the 935 g of water flow from the seawater side to the freshwater side. This process can be observed in Figure 1.7.

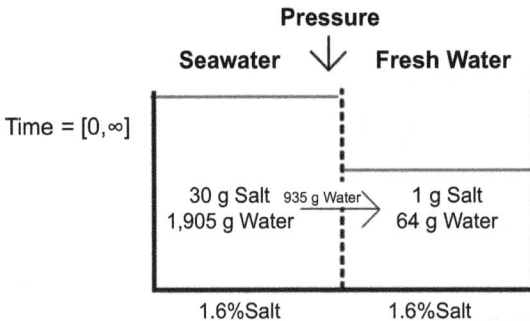

Figure 1.7: Reverse osmosis process.

After applying the same pressure as the osmotic pressure, equilibrium will again be reached, and the initial state will be created. Since typically the goal is to produce fresh water from seawater, and not to prevent osmosis from happening by reaching an equilibrium like the one depicted in Figure 1.8, a pressure greater than the osmotic pressure will need to be applied to produce more fresh water by consuming seawater.

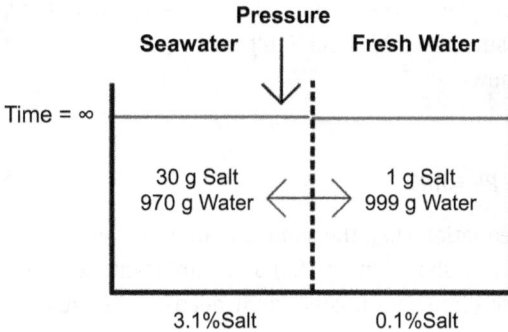

Pressure

Seawater | **Fresh Water**

Time = ∞

| 30 g Salt | 1 g Salt |
| 970 g Water | 999 g Water |

3.1%Salt   0.1%Salt

Figure 1.8: Equilibrium step.

1.5 Equations

RO membranes are defined by a simple set of equations. These equations are used to control the flow of a solvent like water through the membrane, as well as the flow of solutes like salt through the membrane, and also finally to calculate the osmotic pressure needed to start producing fresh water from higher concentration water. Additionally, the equations that are used to characterize an RO system are presented.

1.5.1 Fundamental equations

Osmotic pressure is defined by the Greek letter pi (Π). Osmotic pressure can be calculated by multiplying the temperature (T) in kelvin, with the ideal gas constant (R), the solute concentration (C), and the osmotic pressure coefficient (Φ). This formula is shown in eq. (1.2).

1.5.1.1 Osmotic pressure
The osmotic pressure can be calculated using eq. (1.2). The osmotic pressure coefficient represents how well a solute dissociates in water. For NaCl, which fully dissociates into sodium and chloride ions following eq. (1.1), it will be equal to 1. For other species that do not dissociate at all in water, its value will be equal to 0. Concentration is typically expressed in molar mass (mol/L). The ideal gas law is typically expressed as $0.08314 \ L \cdot bar/K \cdot mol$. Temperature is expressed in kelvin (K):

Osmotic pressure,

$$\pi = \varphi CRT \tag{1.2}$$

One easy way to remember the osmotic pressure equations is by thinking about how, for diluted solutions, the osmotic pressure resembles the Van't Hoff equation for ideal gas laws. The Van't Hoff equation is shown:

Ideal gas law,

$$PV = nRT \tag{1.3}$$

Rearranging the terms of Van't Hoff equation (1.3), the same equation as the osmotic pressure equation (1.2) can be obtained, as shown in eq. (1.4). It is important, however, to highlight that the osmotic pressure dissociation coefficient needs to be factored into this equation:

Ideal gas law,

$$P = \frac{n}{V}RT = CRT \tag{1.4}$$

The net driving pressure (NDP) represents how much of an energy driving force exists across the membrane. This is obtained by discounting the osmotic pressure (Π) from the pressure (P) that is applied to make a membrane permeate water. Pressure units are typically expressed in bar or psi and this formula is given as follows:

Net driving pressure,

$$\text{NDP} = P - \pi \tag{1.5}$$

To better understand how to calculate the osmotic pressure, the following example can be studied. To calculate the osmotic pressure of a solution that has 2 g/L of sodium chloride dissolved in water, eq. (1.2) can be used.

The first step is to transform the mass concentration to molar concentration. This is achieved in the following equation:

Sodium chloride molarity,

$$\frac{2\,\text{g NaCl}}{\text{L}}\frac{1\,\text{mol NaCl}}{58.44\,\text{g NaCl}} = \frac{0.0342\,\text{mol NaCl}}{\text{L}} \tag{1.6}$$

Since sodium chloride fully dissociates in water following eq. (1.1), the osmotic pressure dissociation coefficient can be assumed to be 1. Therefore, it is possible to calculate the concentration of each individual ion dissolved in water as follows:

Sodium chloride dissociation,

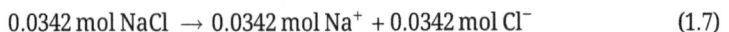

$$0.0342\,\text{mol NaCl} \rightarrow 0.0342\,\text{mol Na}^+ + 0.0342\,\text{mol Cl}^- \tag{1.7}$$

Finally, the contribution of the osmotic pressure needs to be calculated for each individual ion. This is achieved by using eq. (1.2) for each individual sodium and chloride ion. This can be seen in both eqs. (1.8) and (1.9), respectively, where it is evident that both sodium and chloride ions have the same osmotic pressure individual contribution of 0.85 bar each:

Sodium osmotic pressure,

$$P_{Na^+} = \frac{0.0342\,\text{mol Na}^+}{L} \frac{0.08314\,\text{L bar}}{K\,\text{mol}} 298\,K = 0.85\,\text{bar} \tag{1.8}$$

Chloride osmotic pressure,

$$P_{Cl^-} = \frac{0.0342\,\text{mol Cl}^-}{L} \frac{0.08314\,\text{L bar}}{K\,\text{mol}} 298\,K = 0.85\,\text{bar} \tag{1.9}$$

Finally, the total osmotic pressure of sodium chloride can be calculated by adding each individual ion's osmotic pressures. This is shown in eq. (1.10), where it can be seen that the total osmotic pressure of a 2 g/L sodium chloride solution equals 1.7 bar. A rule of thumb to quickly estimate the osmotic pressure is to divide the total dissolved solids by 100. This gives an approximation of the osmotic pressure in psi. To have it in bar, this resulting value needs to be multiplied by 14.5:

Sodium chloride osmotic pressure,

$$P_T = P_{Na^+} + P_{Cl^-} = 1.7\,\text{bar} \tag{1.10}$$

1.5.1.2 Water flux
Water flux across a membrane (F_w) is defined as the multiplication of the water permeability value, called the A-value (A), with the net driving pressure, which is obtained by subtracting the osmotic pressure gradient (Π) from the pressure gradient applied to the membrane (P). The flux of water is typically expressed in US gallons per square feet per day (gfd) or in liters per hour per square meter (LMH). The A-value expresses the membrane water permeability coefficient, and it is typically expressed in US gallons per square feet per day per psi (gfd/psi) or in liters per hour per square meter per bar ($L/m^2 \cdot h \cdot bar$ or simply LMH/bar). Pressure is typically expressed in psi or bar and the formula is

Water flow,

$$F_w = A(P - \pi) \tag{1.11}$$

It is important to notice how this equation resembles Darcy's law equation, which states that a flow across a porous membrane is proportional to the pressure that is applied. Darcy's law can be seen in eq. (1.12), where k represents the permeability coefficient, μ

the dynamic viscosity, and L the membrane thickness. All these parameters can be incorporated into the A-value membrane permeability coefficient:

Darcy's law,

$$F_w = \frac{k}{\mu\,L}P \tag{1.12}$$

This is the same equation that governs the transport of water across an UF membrane. Therefore, it can be concluded that for a solvent like water, when it faces a semipermeable membrane where it can pass freely through it, it acts as a pressure-driven convective flow filtration.

1.5.1.3 Salt transport

The flux of salt across a membrane (F_s) is described as the multiplication of the salt coefficient value, also referred to as B-value, with the concentration gradient of solutes across the membrane. The flux of salt across a membrane is typically expressed in pounds per square feet per day (lbfd) or in grams per hour per square meter (GMH). The B-value, or salt diffusion coefficient, is typically expressed in US gallons per square feet per day (GFD) or in liters per hour per square meter ($L/m^2 \cdot h$ or simply LMH). Concentration is usually expressed in pounds per US gallon (lb/gal) or grams per liter (g/L). This formula is described in the following equation:

Salt flow,

$$F_s = B\;C \tag{1.13}$$

It is important to notice how this equation resembles Fick's law equation. Fick's law describes the diffusion transport of mass across a membrane. As salt cannot pass through the cavities that are created in a membrane, it needs to pass through diffusion. It can be observed how diffusion is noted with a D, and this corresponds to the B-value diffusion coefficient. The concentration gradient stays the same and the formula is given as follows:

Fick's law,

$$F_s = D\;C \tag{1.14}$$

1.5.1.4 The difference between convective and concentration-driven flows

Typically, a pressure-driven convective flow is several orders of magnitude higher than the mass transfer flow that can be achieved through diffusion. This is why RO membranes are able to separate a solute from a solvent so effectively. This mainly happens because, for a semipermeable membrane, a solvent like water can travel across the membranes by following a pressure gradient. This means that water perceives "pores" across the membrane. However, salt cannot pass through these "pores," because due to solvation, dissociated species in water are too large to pass through these "pores," and the only way they can pass through a membrane is through diffusion.

The following examples illustrate the different orders of magnitude difference between a convective flow like water and a diffusion flow like sodium chloride across the membrane.

To calculate the flux of water across a membrane, it can be assumed a water permeability A-value of 4 LMH/bar, a 15 bar feed pressure, and a 1 bar osmotic pressure:

Water flow example,

$$F_w = A(P - \pi) = 4(15 - 1) = 56 \, L/m^2h = 56,000 \, g/m^2h \qquad (1.15)$$

To calculate the flux of salt across this same membrane, a salt diffusion coefficient of 0.2 LMH and a concentration of salts of 2 g/L can be assumed:

Salt flow example,

$$F_s = B \, C = 0.2 \cdot 2 = 0.4 \, g/m^2h \qquad (1.16)$$

As can be observed from this example, RO membranes are really selective to water mass transport across the membrane when compared to salt mass transport. These several orders of magnitude difference in mass transport clearly illustrate the difference between convective and diffusive flow.

1.5.2 System equations

An RO system is mainly composed of a feed flow (Q_f), and then this feed flow gets divided between the filtrated flow that is treated with the membrane active layer, which is called the permeate flow (Q_p), and the concentrate flow (Q_c), which has all the rejected salts or spices that could not permeate the membrane. Concentrate flow is sometimes also referred to as brine or retentate. Flows are usually specified in cubic meters per hour or day (m^3/h or m^3/day), or in US gallons per day (gfd). Plant capacity represents the permeate flow a desalination plant can produce, and it is usually expressed in millions of liters per day (MLD). 1,000 m^3/day equals 1 MLD. A simple RO diagram is found in Figure 1.9.

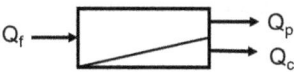

Figure 1.9: Reverse osmosis diagram.

An RO system is characterized by having a close water mass balance. This means that all the water that is entering the RO membrane (Q_f) needs to exit the RO system through either the permeate (Q_p) or through the concentrate flow (Q_c). This formula is depicted as follows:

Feed flow,

$$Q_f = Q_p + Q_c \tag{1.17}$$

An RO system is also characterized by having a neutral salts mass balance. This means that all salts entering the system will also exit the RO membrane through either the permeate side or the concentrate side. Individual salts concentrations for the feed (C_f), permeate (C_p), and concentrate (C_c) are typically represented in g/L or in mg/L (ppm). This is represented as follows:

Salt mass balance,

$$Q_f C_f = Q_p C_p + Q_c C_c \tag{1.18}$$

Flux (F) represents a flow (Q) relative to the membrane active area (A) it is permeating. Flux is typically measured in cubic meters per hour or day per square meter (m^3/h or m^3/day) or in US gallons per day per square feet (US gal/(d · ft^2) or gfd). This formula is given as follows:

Flux,

$$F = \frac{Q}{A} \tag{1.19}$$

RO system recovery (R) represents the process water yield, and is calculated by dividing the permeate flow (Q_p) by the feed flow (Q_f). It is expressed as a percentage. This formula is shown as follows:

Recovery,

$$R = \frac{Q_p}{Q_f} \tag{1.20}$$

Salt passage (SP) represents the percentage concentration of salt that passes through the membrane compared to the initial salt concentration being treated. It is calculated by dividing the concentration of salt in the permeate (C_p) by the concentration of salt in the feed (C_f). This parameter is useful from a physics point of view as it allows for the direct comparison of two different membrane performances. This formula is shown as follows:

Salt passage,

$$SP = \frac{C_p}{C_f}$$ (1.21)

Salt rejection (SR) represents the percentage of how much salt a membrane is rejecting. It is calculated by subtracting 1 minus salt passage (SP). This parameter is useful as it enables one to quickly realize how much solute or salt is being rejected by a membrane system. However, in order to perform a comparative evaluation of the performance of two different membranes, it is usually necessary to do the comparative evaluation using the salt passage parameter. This formula is shown as follows:

Salt rejection,

$$SR = 1 - SP$$ (1.22)

Another factor that is calculated is the plant availability (Av). This represents the amount of time, in percentage, that the plant is in operation producing water (t_{op}) and therefore not stopped, versus the total time of the time period being considered (t_T). This formula is shown as follows:

Availability,

$$Av = \frac{t_{op}}{t_T}$$ (1.23)

1.5.3 Factors affecting membrane performance

Several factors can affect RO membrane performance. The three factors that are usually most relevant are the effect that feed pressure increase, feed concentration increase, and feed temperature increase have on membrane performance. In order to understand how these factors change, only three equations are required. These are the osmotic pressure equation, shown in eq. (1.5), the water transport equation, shown in eq. (1.11), and the salt transport equation, shown in eq. (1.13).

A summary table highlighting these interactions can be found in Table 1.2.

1.5.3.1 Feed pressure
When feed pressure increases, water flux also increases as a result of an increase in the pressure. Salt passage across the membrane stays constant, but since more water is passing across the membrane, whet the water flow is divided by the same amount of salt, the final salt concentration in the permeate decreases. Therefore, salt rejection increases.

Table 1.2: Summary of the effect of feed pressure, concentration, and temperature on the salt rejection and flux of a reverse osmosis membrane.

	Flux	Salt rejection	Equation
Pressure ↑	↑	↑	$Fw = A \cdot (P{\uparrow} - \pi)$
			$Fs = B \cdot C$
Concentration ↑	↓	↓	$Fw = A \cdot (P - \pi{\uparrow})$
			$Fs = B \cdot C{\uparrow}$
Temperature ↑	↑	↓	$Fw = A{\uparrow} \cdot (P - \pi{\uparrow})$
			$Fs = B{\uparrow}{\uparrow} \cdot C$

1.5.3.2 Feed concentration

As feed concentration increases, osmotic pressure also increases. This leads to a direct reduction in the water flux across the membrane, as there is less net driving pressure available for the membrane to permeate. Additionally, as concentration increases, the flux of salt directly increases. This leads to a decrease in water passing through the membrane, which is divided by a higher salt passing through the membrane, thus increasing the salt passage and decreasing the salt rejection.

1.5.3.3 Feed temperature

As feed water temperature increases, the water becomes less viscous. This means that with the same amount of energy, more water can permeate through the membrane. This eventually leads to an improvement in the water permeability value (*A*-value) and therefore the permeate water flux. With regard to salt rejection, an increase in temperature improves the salt diffusion factor (*B*-value) much more. Therefore, there is a higher increase in salt passing through the membrane than water passing through the membrane, and as a result, salt rejection decreases.

1.5.3.4 Concentration polarization

Concentration polarization is the phenomenon in which, as the membrane removes water from the feed solution, solutes are pushed toward the boundary layer of the membrane. This leads to a decrease in performance in the RO membrane system, as the membrane, being a filtration system, sees the concentration in the boundary layer and not in the feed solution. As this salt concentration is higher, it means that the membrane experiences a decline in flux and salt rejection due to this phenomenon [18]. Concentration polarization can be minimized by controlling the membrane recovery rate, as well as through membrane element design elements such as a feed spacer that is able to provide proper mixing and therefore minimizes the accumulation of solutes in the membrane boundary layer.

1.6 Reverse osmosis membranes

RO and NF membranes are pressure-driven membrane filtration processes, where feed pressure greater than the osmotic pressure is used to filtrate water through the membrane.

Commercial elements used in large industrial installations are standardized. They are usually referred to as spiral-wound polyamide-based membranes configured in a cylindrical shape with a typical diameter of 8 inch (20 cm) and a typical length of 40 inch (1 m). For smaller industrial applications where the water capacity required is lower, elements with a 4-inch diameter and 40-inch length are also used. Finally, elements used in home drinking applications are less standardized, and their size in terms of diameter and length can vary depending on each manufacturer. Examples of these elements are 1.8-inch and 2.5-inch diameter elements with 12-inch or 14-inch lengths. An example of a DuPont FilmTec™ membrane like the BW30 PRO-400 element can be found in Figure 1.10. In this example, feed flow will enter the membrane through the anti-telescoping device on the left. The permeate flow will be collected in each membrane leaf and finally collected through the inner permeate water tube. The permeate flow will exit the membrane through the permeate water channel located in the center of the membrane, leaving through the right of the membrane. The concentrate flow will also exit the membrane through the right part via the anti-telescoping device on the right part. It is useful to realize that the remaining feed water that exits the membrane is what is called concentrate.

Figure 1.10: DuPont FilmTec™ BW30 PRO-400 membrane.

An RO membrane is typically composed of an active polyamide base layer that is around 0.2 µm thick. This polyamide membrane is also referred to as the active layer, as it is the one responsible for separating the salt solutes from the water solvent. As this membrane is so thin, in order to be able to precipitate it during the phase inversion process, a support layer typically consisting of polysulfone is used. This allows the proper precipitation of polyamide on the polysulfone. The polysulfone layer has a thickness of around 40 µm. As this thickness is still rather thin, in order to enhance its mechanical stability, this layer is put in a polyester reinforcing layer, typically consisting of a 120 µm thick layer. This three-layer structure is usually referred to as a thin-film composite (TFC) layer [19]. A schematic of this arrangement corresponding to a thin-film composite RO membrane with its multilayer configuration can be found in Figure 1.11.

Polyamide	0.2 µm
Polysulfone	40 µm
Polyester	120 µm

Figure 1.11: Thin-film composite reverse osmosis membrane multi-layer composition.

A scanning electron microscopy (SEM) image, courtesy of DuPont™ FilmTec™ membranes, depicting the main three layers in an RO membrane can be found in Figure 1.12.

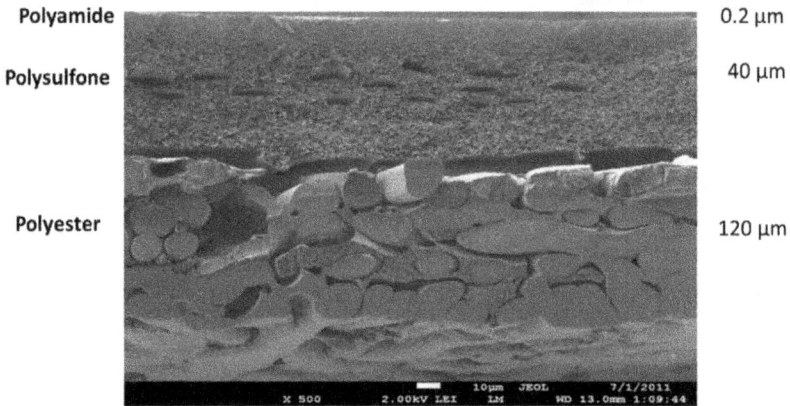

Polyamide	0.2 µm
Polysulfone	40 µm
Polyester	120 µm

Figure 1.12: SEM image of a reverse osmosis membrane.

NF membranes are very similar to RO membranes. Their main difference is that the active layer usually consists of a polypiperazine polymer. Typically, NF membranes are used when only certain solutes need to be separated, but not all of them. This allows for significant energy savings. Examples of their use are sulfate removing NF membranes. These membranes can let sodium chloride pass through their active layer, but they remove sulfates and other divalent ions. This is especially useful for oilfield applications, where seawater is used for injecting into the oil wells. In this application, no sodium chloride needs to be removed. However, to prevent multiple problems, sulfate needs to be removed. By using NF membranes, the operating pressure can be reduced from 70 bar to 15 bar, therefore saving a lot of energy.

Typically, an RO membrane spiral-wound element consists of multiple polyamide sheets that are rolled together. Each membrane sheet is separated from the one on top by a feed spacer. Inside each membrane, there is a permeate spacer. The role of the spacers is to provide mechanical stability to the RO element so that it can be properly folded. The feed spacer also plays a crucial role in minimizing the concentration polarization effect, as well as controlling biofouling and saving energy. Minimizing the dead spaces inside a membrane is important to prevent biological growth inside a membrane. The schematic shown in Figure 1.13 illustrates the main parts of a spiral-wound RO element.

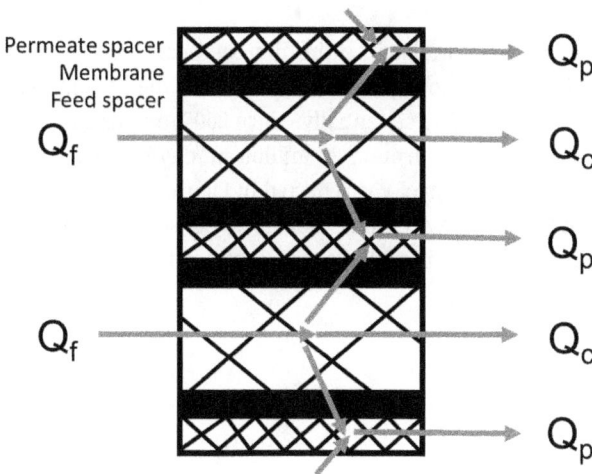

Figure 1.13: A spiral-wound reverse osmosis elements with its parts.

1.6.1 The history of desalination

RO membrane technology was first developed by the University of California, Los Angeles (UCLA) by Dr. Sidney Loeb and Dr. Srinivasa Sourirajan in 1959. The RO membrane that they developed was made of cellulose acetate (CA) [20].

The first commercial RO plant was built in 1965 in Coalinga, California. It was a municipal drinking water plant with a capacity of 5,000 US gallons per day. This plant used the cellulose acetate RO membranes that S. Loeb and S. Sourirajan developed [21].

John E. Cadotte developed the NS-100 membrane, the first non-cellulose acetate membrane. Later, in 1977, J.E. Cadotte developed the NS-101 membrane, the first poly-amide-based thin-film composite (TFC) RO membrane [22].

The development of thin-film composite RO membranes based on polyamide represented a significant breakthrough in the desalination industry. This is due to the superior characteristics of polyamide-based membranes compared to cellulose acetate-based membranes. Polyamide-based membranes offer higher water permeability and higher salt rejection than cellulose-based membranes. Additionally, poly-amide based membranes offer higher resistance to chemicals [23].

RO technology started to be widely adopted after the development of thin-film composite polyamide-based RO membranes. These membranes mainly consist of a thin polyamide layer on top of a porous support layer. This has led to positioning TFC membranes as the most effective technology for desalinating water [21].

1.6.2 The importance of desalination

Since its early inception in the seventies and with thin-film composite membranes being established as the most effective technology to desalinate water, the desalination installed operational capacity has increased exponentially. Since 2005, the desalination installed capacity has increased in a ratio of around 20 million m^3/day every 5 years. The evolution of the desalination installed capacity is reflected in Figure 1.14 [24].

It should be noted that as of 2023, there are more than 16,000 desalination plants in the world. All these plants provide a total installed capacity of newly created fresh-water equivalent to 100,000,000 m^3/day [25].

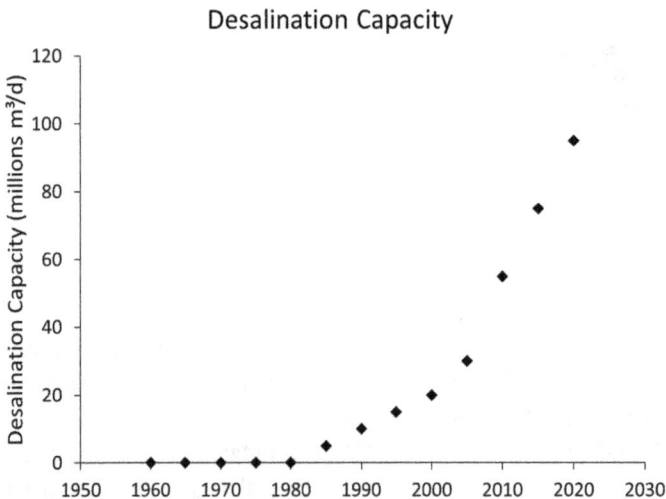

Figure 1.14: Operational desalination capacity.

1.7 Performance monitoring

RO systems are typically designed to provide a constant yearly water production capacity. Therefore, systems are designed at a constant flux rate. So, when systems suffer from fouling or changes in operating conditions such as a temperature decrease or salinity increase, typically their flux rate would decrease. To prevent this, more energy is used so that the membrane system can compensate for the decrease in membrane permeability and be able to provide the same operating flux.

Therefore, it is of utmost importance to periodically monitor the performance of RO systems. There are three key parameters that need to be monitored in RO.

The first one is the energy consumption, monitored through the high-pressure pump operating pressure. This is an important parameter because the energy consumption directly impacts the operating expenses (OPEX) of an RO pump. Additionally, the high-pressure pump needs to have enough capacity to increase the pressure it is delivering to the membrane system, which in turn might affect the capital expenses (CAPEX). If the pump cannot deliver enough pressure, the whole desalination installation can start suffering from a decrease in the water it is delivering. This can lead to water shortages and even being outside the offset contract. This is why typically plants are designed in a conservative way. To properly size the high-pressure pump, typically the lowest yearly temperature is used to size the pump, as the lowest water temperature will provide the highest pressure needed to sustain the targeted design installation permeate flow capacity.

The second parameter that is key to monitoring a membrane system is the permeate water quality. Typically, the water conductivity is monitored. Conductivity is typically expressed in µS/cm. Conductivity is used to estimate the total dissolved solids (TDS) salinity of water. A good rule of thumb for low salinity water is that 2 µS/cm is equal to 1 mg/L (ppm) salinity. For seawater types, a good rule of thumb is that 1.4 µS/cm is equal to 1 mg/L (ppm). Sometimes, besides general salinity measured by conductivity, specific spices that are important are also specified. This can involve measuring specific targets such as alkalinity, boron, and pH, among others. Water quality is of utmost importance because, ultimately, water treatment plants are usually designed to provide a warranted water quality that is typically limited to be below a certain limit. Therefore, water quality typically acts as the independent variable when designing a membrane system. Typically, a membrane installation is designed taking into account the highest water temperature thought the yearly temperature cycle. This is because at the highest temperature is when the water quality will be the worst, and therefore show the highest permeate water total suspended solids value.

The third important parameter to monitor is the membrane pressure drop (dP). Pressure drop is a factor that is important to measure as it directly affects energy consumption. However, this parameter is key because when pressure drop increases, it means that the membrane feed-concentrate channel is getting blocked. This can happen if the membrane is experiencing fouling, as when the membrane gets fouled, it becomes obstructed, making it more difficult for water to travel across the membrane. The key problem this issue presents is that if the membrane gets too blocked, it can start to get mechanically damaged, eventually leading to irreversible damage, and the membrane can stop working as intended. Therefore, when pressure drop starts to increase to higher values, it is a good habit to perform a cleaning in place (CIP) to try to recover the membrane performance to the initial pressure drop values. It should be noted that the higher the pressure drop a membrane has, the more difficult it becomes to clean the membrane and restore its performance to its initial values.

This is why it is so important to clean the membranes early, so that it is easier to recover the membrane performance. Because membrane systems typically suffer from fouling, especially in the areas of the planet where temperature is usually higher and there is therefore a higher water demand due to water scarcity issues, typically plants are designed with redundant trains. A train is a collection of pressure vessels. Each pressure vessel usually contains up to seven membranes in series, and a train usually has dozens of pressure vessels. This means that, for example, in a plant that has 11 RO trains, 1 of these trains can be used to be put in operation when one train needs to go through a chemical cleaning. These designs are called in the industry N–1 designs.

1.8 Normalization

Monitoring feed pressure, permeate conductivity, and pressure drop is of utmost importance in order to be able to anticipate possible problems that might arise from changes in operating conditions and fouling. As was explained previously, as a result of water temperature or water salinity changes, the energy consumption and permeate water quality can change. Therefore, it is of utmost importance to understand if these changes in water quality or energy consumption are due to unexpected problems such as fouling, or if they are normal and expected as a result of the physical principles that were previously explained. This is when membrane normalization comes into play.

The key normalized parameters that need to be analyzed in an RO membrane system are the normalized permeate flow, the normalized salt rejection, and the normalized pressure drop.

Normalization can be done using the equations listed below or using computer-assisted programs such as the FT-Norm PRO that DuPont offers.

1.8.1 Why normalization matters

The example detailed in Table 1.3 is a good example of why normalization matters. This is an example of an RO plant where, after 5 days of operation, the feed water temperature decreases from 25 °C to 21 °C. As can be seen in the monitoring parameters, the plant is delivering a constant permeate production of 50 m³/h with a constant feed flow of 100 m³/h. This represents a 50% recovery. Feed salinity is also constant at 1,000 mg/L. However, it can be observed how feed pressure increases from 8 bar to 8.81 bar, and permeate quality decreases from 30 mg/L to 27 mg/L. Without normalizing the data, it would be challenging to understand if this increase in feed pressure and improvement in water quality is a result of expected thermodynamics, or if it is a result of the membrane getting, for example, fouled. Normalization is the tool that allows us to properly perform this assessment.

In this particular example, it can be seen that normalized permeate flow stays constant, while normalized salt rejection also stays constant. This means that the increase in feed pressure and decrease in permeate total dissolved solids are not related to fouling, but they are related to the normal behavior of the membrane system. This happens, as previously explained, because when the temperature decreases, water quality improves, as less salt passes through the membrane. Additionally, more energy is needed to pump the water through the membrane as its viscosity increases.

If the normalized permeate flow shows, for example, a decrease over time, it could be suspected that fouling is responsible for this loss in membrane permeability. If normalized salt rejection would be, for example, decreasing, this could also indicate the likelihood of issues in the membrane system.

Table 1.3: Normalization example.

Days	Feed flow (m³/h)	Permeate flow (m³/h)	Feed pressure (bar)	Feed TDS (mg/L)	Permeate TDS (mg/L)	Temperature (°C)	Normalized permeate flow (m³/h)	Normalized salt rejection
0	100	50	8.00	1,000	30	25	50	97.0%
1	100	50	8.00	1,000	30	25	50	97.0%
2	100	50	8.19	1,000	29	24	50	97.0%
3	100	50	8.39	1,000	28	23	50	97.0%
4	100	50	8.59	1,000	27	22	50	97.0%
5	100	50	8.81	1,000	27	21	50	97.0%

This normalization example can be easily depicted by observing Figure 1.15, where it can be seen that feed pressure increases, and at the same time, the permeate total dissolved solids decreases. As mentioned previously, this could indicate that the membranes are experiencing fouling.

However, after the data has been normalized and factors such as temperature variation are taken into account, it can be observed in Figure 1.16 that the membranes are operating without any type of fouling. This can be assessed after the feed pressure is normalized to normalized permeate flow, and the permeate total dissolved solids is normalized to normalized salt rejection. Therefore, this increase in feed pressure and decrease in permeate TDS is, in this case, explained by the physical phenomena that

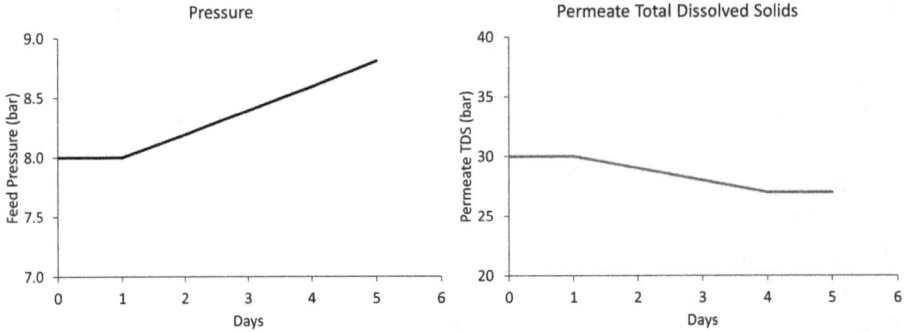

Figure 1.15: Feed pressure and permeate total dissolved solids evolution.

take place when water temperature decreases. This is very important, as it prevents false assessments, where someone could think that there is a fouling issue, while in reality, the membrane is working as expected, and any change in the key operating parameters is due to the physical laws that govern membrane performance, following the equations that have been explained before.

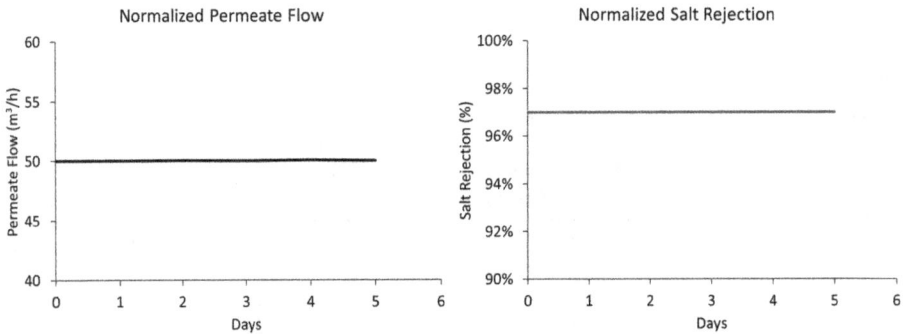

Figure 1.16: Normalized permeate flow and normalized salt rejection evolution.

1.8.2 Equations

Operating data normalization can be done with operating software such as DuPont's FT-Norm PRO software, or by using the equations found in ref. [19].

1.8.2.1 Normalized permeate flow

To normalize the permeate flow (Q_{pn}), the following formula found in eq. (1.24) can be used. It must be noted that the subscript 0 refers to the conditions being normalized to. This can typically be the first data when the system is stabilized, or even an ideal projection that the system is designed to operate at under normal or ideal conditions. The other data without a subscript refers to the data at the time instance being normalized to. Units are in bar for the pressure drop, and in m^3/h or m^3/day for the flows:

Normalized permeate flow,

$$Q_{pn} = Q_p \frac{NDP_0}{NDP} \frac{TCF_0}{TCF} \qquad (1.24)$$

Net driving pressure (NDP) is calculated by taking the feed pressure and subtracting half the pressure drop or differential pressure (dP), the permeate pressure (P_p), and the average feed concentrate osmotic pressure (Π_{fc}). All pressures are measured in bar. This is shown in eq. (1.25). It should be noted that the osmotic pressure in the permeate has been omitted for simplification:

Net driving pressure,

$$NDP = P_f - \frac{dP}{2} - P_p - \pi_{fc} \qquad (1.25)$$

Pressure drop (dP) is calculated using eq. (1.26). To calculate it, the concentrate pressure (P_c) needs to be subtracted from the feed pressure (P_f):

Pressure drop,

$$dP = P_f - P_c \qquad (1.26)$$

The average feed-concentrate osmotic pressure (Π_{fc}) can be calculated using eq. (1.27) [26]. It should be noted that temperature (T) is in kelvin:

Feed-concentrate osmotic pressure,

$$\pi_{fc} = 0.00265 \cdot C_{fc} \cdot T \qquad (1.27)$$

The average feed-concentrate concentration (C_{fc}) can be calculated using eq. (1.28). It uses the feed concentration (C_f) and the membrane system recovery (R). Its units are in mg/L:

Feed-concentrate concentration,

$$C_{fc} = C_f \frac{\ln \frac{1}{1-R}}{R} \qquad (1.28)$$

To calculate the temperature correction factor (TCF), eq. (1.29) can be used. It should be noted that the temperature (T) also needs to be in kelvin. Additionally, the k parameter depends on the water temperature. If the temperature is ≥ 25 °C, $k = 2,640$. If the temperature is ≤ 25 °C, $k = 3,020$:

Temperature correction factor,

$$\text{TCF} = e^{k\left(\frac{1}{298} - \frac{1}{T}\right)}$$

(1.29)

1.8.2.2 Normalized salt rejection

In order to normalize the salt passage (SP_n), the following formula is used. This is displayed as follows [26]:

Normalized salt passage,

$$SP_n = SP \frac{Q_p \cdot TCF_0 \cdot C_{fc0} \cdot C_f}{Q_{p0} \cdot TCF \cdot C_{fc} \cdot C_{f0}}$$

(1.30)

To normalize the salt rejection (SR_n), the formula displayed in eq. (1.31) should be used:

Normalized salt rejection,

$$SR_n = 1 - SP_n$$

(1.31)

1.8.2.3 Normalized pressure drop

The normalized pressure drop (dP_n) can be calculated using the formula displayed in eq. (1.32) [27]. To perform this calculation, the standard pressure drop (dP) is used, together with the division of the baseline average feed-concentrate flow (Q_{fc0}) and the average feed-concentrate flow (Q_{fc}). Units are in bar for the pressure drop, and in m³/h or m³/day for the flows. The viscosity of water (ρ_{fc}) can also be factored in, although in most cases it can be neglected as the differences should be very small. The equation also takes into consideration the dynamic viscosity of water (μ_{fc}):

Normalized pressure drop,

$$dP_n = dP\left(\frac{Q_{fc0}}{Q_{fc}}\right)^{b+2} \left(\frac{\rho_{fc0}}{\rho_{fc}}\right)^{b+1} \left(\frac{\mu_{fc}}{\mu_{fc0}}\right)^{b}$$

(1.32)

The dynamic viscosity of water (μ_{fc}) can be approximated using eq. (1.33) [28], and it is measured in Pa · s or N · s/m². It is also worth noting that the temperature needs to be in kelvin:

Water dynamic viscosity,

$$\mu = 2.414 \times 10^{-5} \times 10^{\left(\frac{247.8}{T - 140}\right)}$$

(1.33)

The exponential constant (b) is an empirical parameter obtained experimentally by fitting the friction parameter of the membrane (f) with its Reynolds number, following eq. (1.34). It should be noted that, for most cases, b can be assumed to be −0.5 [27]:

Friction parameter,

$$f = a \ Re^b \tag{1.34}$$

1.9 Fouling

Fouling is the phenomenon in which membranes get dirty. Fouling is one of the major challenges that is currently affecting the RO industry [29]. The key problem that fouling presents is that it is difficult to foresee, and it is also difficult to deal with once fouling starts to appear. It is sometimes also difficult to identify the type of fouling present.

Fouling is typically characterized by the accumulation of unwanted spices that get deposited inside the membrane, leading to a decrease in its performance. Typically, fouling leads to higher operating costs in membrane systems, as it can result in higher energy consumption, lower water quality, lower water production, higher chemical costs due to cleanings, and even irreversible damage to the membrane. Fouling can be characterized into four major types, plus a fifth one that relates to membrane integrity failure.

1.9.1 Biofouling

The main type of fouling is biological fouling, or simply biofouling. Biofouling occurs when certain bacteria, which have evolved to form a biofilm when they find a solid surface, meet the membrane and the feed spacer. When this happens, these bacteria attach to the membrane and start to form a biofilm.

Biofouling is characterized by an exponential increase in the feed-concentrate pressure drop of the RO membrane. It can lead to an increase in energy consumption, and if not dealt with properly, it can mechanically damage the RO membrane. Biofouling is typically present in the first elements of a pressure vessel.

Once a biofilm is established on the membrane, it is difficult to remove. Useful strategies involve cleaning the membranes with caustic cleanings (CIP). These cleanings are most effective at higher temperatures and pH levels. If possible, based on the membrane characteristics, effective cleanings will involve a pH around 13 at a temperature of approximately 35 °C using NaOH. It is important to allow enough time for the chemicals to soak into the membranes, and then flush the soaked solution effectively. This combination of steps can be repeated multiple times until the soaking solution appears clean.

Other, more preventive approaches involve starving the bacteria from growing. This can be achieved using pretreatment technologies like the DuPont™ B-Free™, which is able to eliminate nutrients before they reach the RO system [30]. This system has been proven extremely effective in preventing biofouling development in RO systems.

Other concepts that are practiced involve using non-oxidizing biocides, such as DBNPA or 2,2-dibromo-3-nitrilopropionamide (DBNPA). This biocide is described as non-oxidizing to the polyamide active layer of the RO membrane. Some considerations that need to be taken into account with this solution are its limited compatibility with drinking water applications and the fact that bacteria can get used to the biocide, and at a certain point, it can stop being effective, allowing biofouling to develop again.

1.9.2 Organic fouling

The second type of fouling is organic fouling. Organic fouling happens when organic molecules accumulate on the membrane surface. It typically leads to a decline in normalized permeate flow.

The main way to deal with organic fouling is through performing caustic-based chemical cleanings (CIP), such as using NaOH. These cleanings are most effective at higher temperatures and pH levels. The same cleaning procedure as with biofouling can be followed.

1.9.3 Particulate fouling

Particulate fouling, or colloidal fouling, refers to the accumulation of particles in the membrane. This can be due to an inadequate pretreatment. This type of fouling leads to a rapid increase in the feed-concentrate pressure drop, as the feed spacer channel fills quickly with particles.

In order to clean particular fouling effectively, it might be necessary to perform caustic-based chemical cleanings (CIP). These cleanings are most effective when the temperature and the pH are higher. If possible, based on the membrane characteristics, effective cleanings will involve a pH around 13 at a temperature of approximately 35 °C.

A good solution to deal with this type of fouling can involve upgrading the pretreatment from a conventional one to a membrane-based one, like UF. Another solution could be ensuring there are no fiber breakages in the UF part, and if there are, repairing them with glue and pins, or replacing the damaged modules with new ones.

1.9.4 Scaling

Scaling, also known as inorganic fouling, is a type of fouling that typically occurs when not too soluble salts start to precipitate on the membrane module. This can happen if water recovery is too high, or if temperature or water composition changes. A recommended way to prevent scaling is consulting a specialized antiscalant company. They have powerful software that simulates the operating conditions at the target recoveries and temperatures and recommends the best antiscalants and their concentrations to use, based on the spices that have a higher risk of precipitating as they risk surpassing their solubility limit. Scaling typically happens in the last elements of a pressure vessel, as that is where there is less water, and the water is more concentrated. So dissolved species have a higher risk of precipitating. Scaling typically leads to an increase in pressure drop, as well as a decrease in water quality or salt rejection. As salts start to precipitate on the membrane surface, this affects the concentration polarization and increases the effective concentration of salts in the boundary layer. Scaling can also therefore lead to a decrease in normalized permeate flow, since the osmotic pressure is greatly increased, thus reducing the net driving pressure.

In order to clean a scaled membrane, it is important, if possible, to autopsy a membrane to understand the type of scaling present. There are multiple companies that specialize in offering this service. To clean the membrane, it is recommended to perform an acid chemical cleaning (CIP). If the membrane characteristics allow it, a pH of 1 at room temperature is effective. It is recommended to use HCl to prevent any further scaling caused by other species such as sulfuric acid. Sometimes, membranes can have some organic or biofouling too. Therefore, it is also recommended to always start with a caustic cleaning as previously described, and then follow the acid cleaning step.

1.9.5 Integrity failure

Integrity failure occurs when the membrane suffers from non-fouling-related damage.

This can involve chemical oxidation of the membrane. This can happen if, for example, sodium hypochlorite (NaOCl) from the UF pretreatment manages to reach the RO membrane, as the polyamide can get damaged and eventually eliminated. Symptoms involve an increase in the normalized salt passage, as since the membrane does not have an active layer, it stops separating salt from water. Once oxidation is detected, it is important to eliminate the source that is causing the chemical leading to oxidation to leach. A strategy to properly address the leaching of NaOCl in UF membranes is described here [31].

Other types of mechanical failures might involve o-ring failure. This can happen when the O-ring that is used to separate a membrane gets pinched, causing a by-pass of water from the feed side to the permeate side. This leads to an increase in the normalized salt passage. In order to fix this, a probing test needs to be done in each membrane connection inside a pressure vessel. A methodology to perform this test is described elsewhere [26].

Compaction can happen when a membrane is operating at too high a pressure and temperature. Compaction can be reversible or irreversible, or a combination of both. Typical effects of compaction involve a high increase in energy consumption, observed by a decline in normalized permeate flow. Compaction can lead to an improvement in water quality, as the membrane becomes denser, making it more difficult for the salt to pass through it. When compaction is identified, it is important to either use compaction-resistant RO membranes or to decrease, if possible, the target permeate flux, especially in periods of high temperature, with the aim to reduce the operating pressure.

References

[1]   R. W. Baker. Membrane Technology and Applications, 3rd edition, Wiley. (2012).

[2]   P. M. Budd, N. B. McKeown, B. S. Ghanem, K. J. Msayib, D. Fritsch, L. Starannikova, N. Belov, O. Sanfirova, Y. Yampolskii, V. Shantarovich. Gas permeation parameters and other physicochemical properties of a polymer of intrinsic microporosity: Polybenzodioxane PIM-1. Journal of Membrane Science and Technology, 325 (2008) 851–860.

[3]   P. Burba, B. Aster, T. Nifant'eva, V. Shkinev, B. Y. Spivakov. Membrane filtration studies of aquatic humic substances and their metal species: A concise overview: Part 1. Analytical fractionation by means of sequential-stage ultrafiltration. Talanta, 45 (1998) 977–988.

[4]   L. V. Saboyainsta, J. L. Maubois. Current developments of microfiltration technology in the dairy industry. Le Lait, 80 (2000) 541–553.

[5]   Degrémont (Suez). Water Treatment Handbook, Lavoisier Publishing. (2007).

[6]   C. Fritzmann, J. Löwenberg, T. Wintgens, T. Melin. State-of-the-art of reverse osmosis desalination. Desalination, 216(1) (2007) 1–76.

[7]   R. Valavala, J. Sohn, J. Han, N. Her, Y. Yoon. Pretreatment in reverse osmosis seawater desalination. A Short Review, Environmental Engineering Research, 16(4) (2011) 205–221.

[8]   L. F. Greenlee, D. F. Lawler, B. D. Freeman, B. Marrot, P. Moulin. Reverse osmosis desalination: Water sources, technology, and today's challenges. Water Research, 43(9) (2009) 2317–2348.

[9]   W. Guo, H. H. Ngo, J. Li. A mini-review on membrane fouling. Bioresource Technology, 122 (2012) 27–34.

[10]   T. Carroll, S. King, S. R. Gray, B. A. Bolto, N. A. Booker. The fouling of microfiltration membranes by NOM after coagulation treatment. Water Research, 34(11) (2000) 2861–2868.

[11]   M. Kumar, T. Culp, Y. Shen. Water desalination: History, advances, and challenges. In Frontiers of Engineering: Reports on Leading-Edge Engineering from the 2016 Symposium, National Academies Press, (2017) 55–132.

[12]   J. Kim, K. Park, D. R. Yang, S. Hong. A comprehensive review of energy consumption of seawater reverse osmosis desalination plants. Applied Energy, 254 (2019) 113652.

[13]   E. Kadaj, R. Bosleman. Energy recovery devices in membrane desalination processes. In Renewable Energy Powered Desalination Handbook. Butterworth-Heinemann, (2018) 415–44.

[14]   Sustainable clean water through solar-powered desalination for water-scarce islands and coastal regions (SDG: 2, 3, 6, 8, 11, 12, 14), Section #SDGAction42477, https://sdgs.un.org/partnerships/sustainable-clean-water-through-solar-powered-desalination-water-scarce-islands-and, consulted on August 19, 2023.

[15]   Y. Fernández-Torquemada, A. Carratalá, J. L. Sánchez Lizaso. Impact of brine on the marine environment and how it can be reduced, Desalination and water treatment, 167 (2019) 27–37.

[16]   S. Casas, C. Aladjem, E. Larrotcha, O. Gibert, C. Valderrama, J. L. Cortina. Valorisation of Ca and Mg by-products from mining and seawater desalination brines for water treatment applications. Journal of Chemical Technology and Biotechnology, 89(6) (2014) 872–883.

[17] L. Wang, J. He, M. Heiranian, H. Fan, L. Song, Y. Li, M. Elimelech. Water transport in reverse osmosis membranes is governed by pore flow, not a solution-diffusion mechanism. Science Advances, 9(15) (2023) 1–12.

[18] S. S. Sablani, M. F. A. Goosen, R. Al-Belushi, M. Wilf. Concentration polarization in ultrafiltration and reverse osmosis: A critical review. Desalination, 141(3) (2001) 269–289.

[19] DuPont™. FilmTec™ Reverse Osmosis Membranes Technical Manual, (2023).

[20] J. Glater. The early history of reverse osmosis membrane development. Desalination, 117(1–3) (1998) 297–309.

[21] M. Heiranian, H. Fan, L. Wang, X. Lu, M. Elimelech. Mechanisms and models for water transport in reverse osmosis membranes: History, critical assessment, and recent developments. Chemical Society Reviews, 52(24) (2023) 8455–8480.

[22] J. E. Cadotte, R. J. Petersen. Thin-film composite reverse-osmosis membranes: Origin, development, and recent advances, synthetic membranes. Chapter, 21 (1981) 305–326.

[23] R. J. Petersen. Composite reverse osmosis and nanofiltration membranes. Journal of Membrane Science, 83(1) (1993) 81–150.

[24] B. Zolghadr-Asli, N. McIntyre, S. Djordjević, R. Farmani, L. Pagliero. A closer look at the history of the desalination industry: The evolution of the practice of desalination through the course of time. Water Supply, 23(6) (2023) 2517–2526.

[25] Climate ADAPT, consulted on August 19, 2023, https://climate-adapt.eea.europa.eu/en/metadata/adaptation-options/desalinisation

[26] DuPont. FilmTec™ FT-Norm software. 2021.

[27] J. E. Johnson, K. Majamaa, C. Yang. Application-driven selection of feed spacers for reverse osmosis. In The International Desalination Association World Congress on Desalination and Water Reuse, vol. 201 (2013).

[28] R. Esteves, N. Onukwuba, B. Dikici. Determination of surfactant solution viscosities with a rotational viscometer. Beyond: Undergraduate Research Journal, 1(1) (2016) 2.

[29] J. Kucera. Reverse Osmosis: Industrial Applications and Processes, Wiley. (2011).

[30] G. Massons, G. Gilabert-Oriol, S. Arenas-Urrera, J. Pordomingo, J. C. González-Bauzá, E. Gasia, M. Slagt. Industrial scale pilot at Maspalomas I desalination plant demonstrates the efficiency of DuPont™ B-Free™ pretreatment–a new breakthrough solution against biofouling. Desalination & Water Treatment, 259 (2022) 261–265.

[31] G. Gilabert-Oriol. Ultrafiltration Membrane Cleaning Processes: Optimization in Seawater Desalination Plants, Walter de Gruyter GmbH & Co KG. (2021) Jun 8.

Chapter 2
Biofouling fundamentals

This chapter explains the fundamentals of organic and biological fouling. Biofouling is characterized by an initial phase of a flat pressure drop increase over time, followed by a second phase consisting of an exponential growth. Organic fouling is characterized by a sudden loss in normalized permeate flow, that later tends to stabilize and plateau. This phase is typically characterized by a flat pressure drop if no biofouling happens at the same time. Special emphasis is put into understanding which factors influence the development of biofouling the most, together with explaining the strategies that can be taken to prevent it. This is illustrated by the concept of the biofouling triangle, where three key factors in biofouling development are identified. These factors are temperature, which promotes bacterial growth; bacteria availability, which is needed in order for bacteria to reproduce; and nutrients availability, which are needed for bacteria to grow and reproduce, and comprises assimilable organic nutrients such as carbon, nitrogen, and phosphorous, together with oxygen, which is used for bacteria to grow.

2.1 Biofouling as the key challenge

Water scarcity is being recognized as one of the main threats that mankind is facing globally [1]. Reverse osmosis (RO) membrane technology has developed as a promising technology to address this problem, holding roughly 44% market share and growing, among all the desalinating technologies [2]. This increase has been possible as materials are improved and costs dropped [3].

Despite the multiple attempts to tackle biofouling, it still remains a key unsolved problem in the water treatment industry [4]. It is currently reported that around 70% of the RO water treatment plants in the Middle East are experiencing problems with biofouling [5]. Additionally, it is also reported that around 83% of all the surface water plants in the United States of America are reported to have problems with biofouling [6]. These facts position biofouling as the key challenge to address in the RO systems that deal with water where bacteria live, reproduce, and eventually form a biofilm.

Moreover, a comprehensive study done by Genesys, a company specializing in performing autopsies to the RO membranes, after studying almost 100 RO seawater membranes that they autopsied and analyzed between 2002 and 2010, revealed that from all the symptoms that those membranes revealed, the most prevalent case among all was biofouling. So, biofouling was identifies as the number one consequence as why membranes where failing. And in particular, biofouling was shown to be prevalent in 27% of all the cases analyzed [7]. In order to corroborate these find-

https://doi.org/10.1515/9783111639079-002

ings and understand if still one decade later, biofouling was still the main reason that accounted for failure of RO membranes, different customers were interviewed globally by DuPont Water Solutions in 2019. As a result of these interviews, it was observed that biofouling was still the key reason for RO membrane failure. In particular, it was the main reason for membranes to fail, with 32% of the cases being accounted showing biofouling [8]. These results are summarized in Figure 2.1, with the main conclusion being that biofouling still represents the key challenge to be solved in the RO water treatment industry.

Main reason for reverse osmosis membranes failure

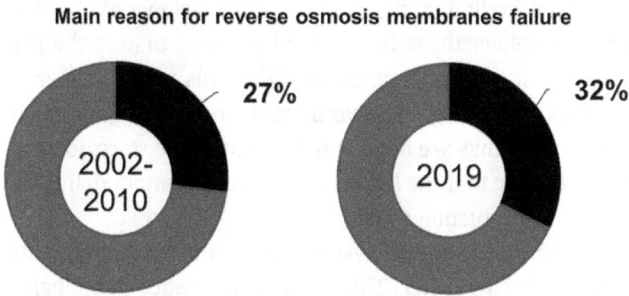

Figure 2.1: Biofouling has been and still is the main problem.

2.2 Fouling in reverse osmosis membranes

Fouling in RO elements takes on many forms. These are typically categorized as inorganic scaling, colloidal or particle fouling, organic fouling, and biological fouling [9]. The first two are generally solved through advanced pretreatment technologies to soften the water like lime softening, antiscalant dosing, or ultrafiltration pretreatment to remove suspended solids. There are also pretreatment technologies to reduce the concentration of dissolved organic material from thousands of ppm down to 40–60 ppm, but reducing the concentration further is less efficient and can be costly [10]. Because of this, the RO systems are expected to share the burden and are often exposed to waters with concentration of organic matter >10 ppm. Consequentially, they suffer from organic fouling and biological fouling.

2.2.1 Organic fouling and its consequences

Organic fouling is defined as the accumulation of organic contaminants on the membrane surface [11]. This accumulation creates a drop in the effective membrane permeability, which lowers the membrane flux and increases the energy of operation [12].

It should be noticed that an organically fouled membrane can undergo a chemical cleaning (CIP) in order to attempt to restore the membrane original permeability and salt rejection. The cleaning procedures are described in this book in detail [13]. The cleaning can either fully recover the original water permeability and salt rejection of the membrane, or it can just recover it partially. Therefore, it is always important to design the high-pressure pumps, so that they can deliver more pressure at the required flow, which can compensate for the increased energy consumption that the RO system will require, as it gets progressively organically fouled.

It is important to take into account that while organic fouling can take a toll on the energy consumption of an RO installation, the worst-case scenario would be that, if the high-pressure pump cannot deliver the sufficient feed pressure to keep the permeate flow constant, the system can reduce its permeate flow. This might happen in the case that either the cleanings stop being effective in restoring the original membrane permeability, or also if no cleanings are done and the membrane starts to experience such a degree of organic fouling that the high-pressure pump cannot deliver its required pressure to keep the flow production constant.

However, it should be noticed that if the RO system starts to lose water production, the installation will keep producing water. This is a key difference in biological fouling, as with biofouling, if the RO system keeps experiencing biofouling and no cleanings are done, or the cleanings are not effective in restoring the membrane's required performance, the high pressure drop might lead to a mechanical collapse of the RO membrane element; this can potentially trigger an installation shutdown, as in the worst case scenario, the membranes would be mechanically damaged and they might need a replacement.

2.2.2 Biofouling and its consequences

Biofouling is defined as the growth and accumulation of microorganisms and bacteria, and the agglomeration of extracellular materials on the solid–liquid surface within the feed channel of a spiral-wound reverse osmosis membrane [14]. The extracellular polymeric substances (EPS) films are especially troublesome to clean. The films anchor on surfaces with low shear and form webs within the feed spacer architecture, as it can be seen in Figure 2.3. This "web" creates high resistance to water flow through the feed channel of the element and displays as an increase in feed-concentrate pressure drop (dP) across the RO pressure vessel. High dP leads to hydraulic imbalance and can result in module damage. Additionally, like organic fouling, biofilms can affect feed channel transport properties and reduce the effective membrane permeability. Both system dP increase and drop in permeability increase the energy of operation but also lead to frequent cleanings to regain element performance. In total, fouling affects energy consumption, element lifetime, water productivity, and cost of water produced [15].

This behavior is summarized in Figure 2.2. In this diagram it can be observed how biofouling leads to a pressure drop increase, which leads to a higher energy consumption. This increase in differential pressure later leads to a chemical cleaning (CIP), which involves the use of chemicals. This chemical cleaning is performed in an attempt to restore the initial performance of the RO membrane before it begins to experience biofouling. However, each cleaning leads to a decrease in the uptime, as the installation needs to be stopped in order to clean the membrane. Biofouling can also lead to a deterioration on the permeate water quality. This phenomenon is explained in the bibliography as the biofilm enhanced osmotic pressure (BEOP). This happens as the biofilm accumulation on the membrane surface tends to accumulate salt on it, thus leading to an increase in the concentration polarization [16, 17]. Finally, after multiple chemicals cleanings on a membrane, and after repeated exposure to a growing biofilm, the membrane can experience a loss in its performance or some mechanical damage, which can lead to a premature replacement of the RO membrane, thus shortening its lifetime. All this can significantly affect the total cost of water in an RO water treatment plant.

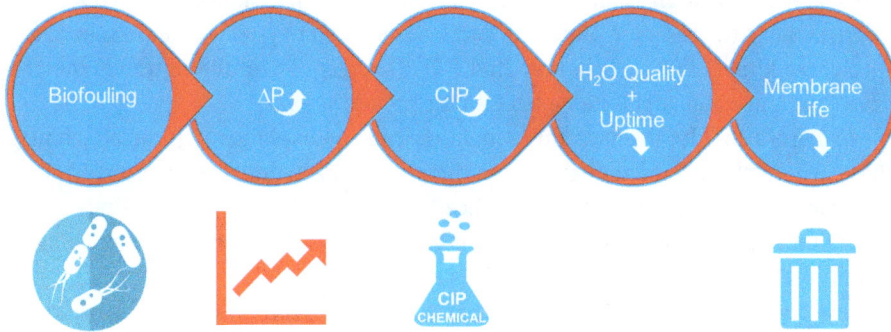

Figure 2.2: Impact of biofouling.

It should be noticed that when compared to organic fouling, biofouling presents a higher thread to the RO system. This higher risk is explained by the unique consequences of each type of fouling. So, if a membrane is suffering from organic fouling, the worst thing that can happen to an RO system is that the high-pressure pump cannot deliver enough pressure to keep the permeate flow constant, and the installation can lose water production capacity. On the other hand, when an RO system is suffering from biofouling, and when either cleaning is not performed or the cleaning performed is not enough to operate at a lower pressure drop, the membrane can collapse because of an excessive differential pressure drop force across the feed-concentrate channel. This collapse might lead to a mechanical failure of the RO membranes. This might require stopping the RO installation in order to replace the damaged membranes, causing a significant downtime and leading to higher costs.

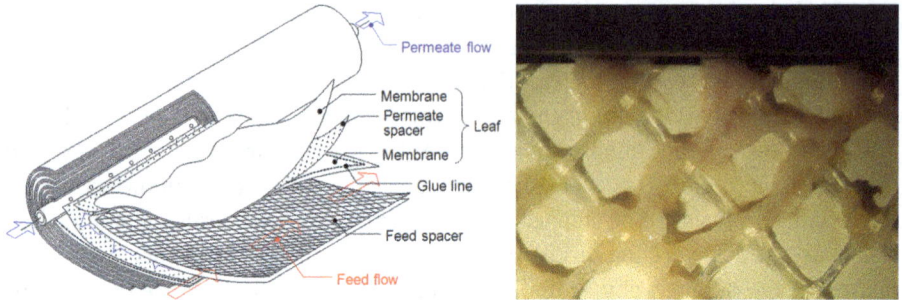

Figure 2.3: Reverse osmosis element configuration (left), biofouled feed spacer (right).

2.2.3 The role of the biofilm in biofouling

Biofouling is caused by the growth of a biofilm. A biofilm is characterized by microbial cells that are attached to a surface and are bound together by the EPS that the cells segregate themselves and glue on to the surface [18]. It should be noted that microorganisms are present in almost all surfaces and water systems and are specialized to colonize almost any surface they find [19]. They grow with the nutrients they receive from the water that is being fed into the membrane [20].

Usually, a biofilm presents three main characteristic phases [21]: the attachment phase, where organic matter deposits on the membrane and serves as a support layer for bacteria to start attaching to the surface; the growth phase, where biofilm starts to grow; and the dispersal phase, where the biofilm is mature and starts detaching as part of it is too big to keep attached to the surface.

It is reported that a biofilm consists mostly of EPS, with an EPS concentration of 99.6–99.9% of its total biofilm volume, leaving 0.1–0.4% of the remaining volume to the bacteria [20].

A biofilm is mostly composed of water, and therefore, resembles water. It is reported that the EPS consists of highly hydrated biopolymer that form a hydrogel, with a water content of 95–99% [22].

2.3 The causes of biofouling

As shown in the biofouling triangle, for biofouling to develop into a solid surface, it needs to be bacteria, nutrients, and temperature. These concepts are described below.

2.3.1 The biofouling triangle

The propensity of a spiral-wound reverse osmosis membrane to develop biofouling can be exemplified through the biofouling triangle that is shown in Figure 2.4. The biofouling triangle is schematic, inspired by the fire triangle, to show the three key aspects that are needed for biofouling to develop into a membrane system. These three aspects are bacteria, nutrients, and temperature. It should also be noted that for a biofilm to develop, a substrate or solid surface is needed.

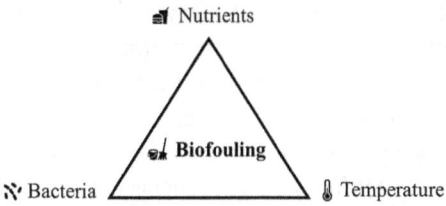

Figure 2.4: The biofouling triangle.

2.3.2 Bacteria

Firstly, bacteria are needed. Without bacteria, no biofilm can develop. Bacteria are everywhere, but for a biofilm to form there needs to be bacteria that are capable of building a biofilm. Getting rid of bacteria is extremely difficult, as despite having a treatment that can eliminate the majority of bacteria, there will always be some bacteria that will not be removed. This is of utmost importance, as one single bacterium that is passing can adhere to the RO membrane and start building a biofilm. Another important factor is the fact that even if there is a bacterial removal technology, as the downstream pipes and equipment are not sterilized, there will always be bacteria available to start forming a biofilm when the conditions are met.

There are several solutions that can be used to reduce the bacteria load to an RO membrane. However, for an RO system, it should be noted that sodium hypochlorite, also referred as chlorine or bleach, cannot be used, as it would oxidize the polyamide of the reverse osmosis membrane. Chlorine is vastly used in upstream ultrafiltration pretreatment systems, as it is a very effective biocide and organics cleaner. Another solution that has been proposed is the used of nonoxidizing biocides, such as 2,2-dibromo-3-nitrilopropionamide (DBNPA) [24]. Chloramines have also been proposed as a way to reduce the bacterial load to an RO system. However, it should be noticed that depending on the specifics of each system, this could lead to a deterioration of performance of the RO system [25].

This book will delve into the use of nonoxidation biocides such as DBNA and its role in mitigating biofouling.

2.3.3 Nutrients

Nutrients are essential for bacteria to live, reproduce, and contribute to the effort of developing a biofilm. The main essential nutrients that bacteria need to reproduce are a source of organic assimilable carbon (AOC), nitrogen (N), phosphorous (P), and oxygen. It should be noted that oxygen is also included as a nutrient, since it meets the *Oxford English Dictionary* definition of a nutrient, which is defined as a substance that provides nourishment for the maintenance of life and for growth [26].

There are several solutions that have been proposed to limit nutrients availability, and the author believes this might be the most successful strategy to prevent biofouling in reverse osmosis system. Some authors have focused on limiting the availability of phosphate, since typically phosphate is found in µg/L (ppb) in natural waters, while other nutrients such as nitrate and carbon are found on the mg/L (ppm) ranges [27, 28]. Other strategies involve adding a pretreatment system between the RO membrane and the ultrafiltration pretreatment that focuses into reducing the amount of nutrients available for the bacteria to digest, such as the DuPont™ B-Free™ system [29, 30].

This book will explain this system and discuss its effect in mitigating biofouling in RO systems.

2.3.4 Temperature

Temperature is a key parameter for biofilms to develop in RO membranes. It has been shown that at higher water temperatures, biofilm development and pressure drop increase in a membrane system [31].

Reducing water temperature has a clear effect on limiting the growth of biofouling in RO systems. However, limiting water temperature as a strategy to reduce biofouling does not seem practical, especially because to achieve this goal, a large cooler with a high energy expense will be required.

This book will show the impact that water temperature can have on limiting biofouling development.

2.4 Identifying biofouling and organic fouling

2.4.1 No fouling

The ideal situation where not fouling is developing is shown in Figure 2.5. In this case, the water permeability (*A*-value), represented in the diagram by an "F", remains flat over time, as there is no organics or fouling matter that deposits on the membrane that would lead into a decrease in the water permeability. At the same time, the salt rejection (SR), which equals one minus the salt passage (*B*-value), remains constant, as there is not additional fouling layer that would make more difficult for salt to pass through the membrane. Also, the pressure drop remains flat, as there is no obstruction of the feed-concentrate channel in the reverse osmosis membrane element that would lead to an increase on the pressure drop [32, 33]. This happens as no particulate matter or biofouling grows on the membrane.

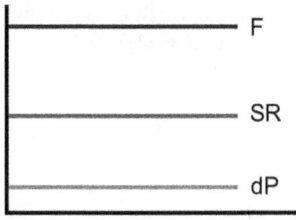

F

SR

dP

t   **Figure 2.5:** No fouling typical behavior.

2.4.2 Biofouling

Biofouling is characterized by a sudden increase of pressure drop. This typically happens after the RO membrane has been in operation for some time, and it is a common occurrence after the organic fouling period has already started. This increase in pressure drop is the result of the biofilm that is growing in the RO membrane, which leads to a blockage of the feed-concentrate channel where the water is flowing through the RO membrane. As pressure drop increases, there is less available feed pressure for the water to permeate through the membrane. This typically leads to a decrease on the normalized water permeability (*A*-value), which also leads to a decrease in salt rejection as there is less net driving pressure [32, 33]. This decrease in salt rejection can also be explained through the accumulation of salts on the membrane surface due to the retention of these salts by the biofilm. This phenomenon is called biofilm enhanced osmotic pressure (BEOP) [16, 17]. This typical behavior of biofouling is exemplified in Figure 2.6.

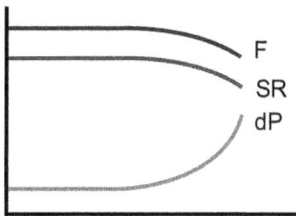

F
SR
dP

t   **Figure 2.6:** Biofouling diagram.

2.4.3 Organic fouling

Organic fouling is usually the first to happen. During this fouling type, organics quickly absorb on the membrane surface, leading to a quick decrease in water permeability (A-value, here noted as flux), as well as a typical improvement in the salt rejection (SR) [32, 33]. This usually happens as the membrane gets fouled; it is more difficult for water to pass through it, but it is also difficult for salt to pass through the membrane. During this period, pressure drops tend to be constant, as there is no biofilm blocking its feed-concentrate channel. This typical behavior of biofouling is exemplified in Figure 2.7.

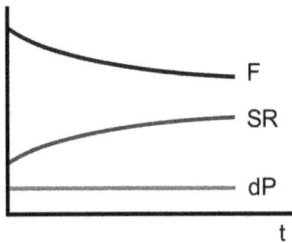

Figure 2.7: Organic fouling typical behavior.

2.5 Identifying biofouling and organic fouling in an example

This is an example of real experiment that is explained in detail in this book, where the first period is characterized by organic fouling developing in the RO membrane. This is followed by a second period where biofouling starts to grow and develop over an already organically fouled RO membrane.

2.5.1 Pressure drop

As it is explained above, the organic fouling period can be quickly identified by having a flat or almost flat pressure drop evolution over time. After some time in operation, the increase of pressure drop over time that corresponds to the biological fouling period can already be noticed. These two different fouling types can be identified in Figure 2.8.

2.5.2 Water permeability

During the organic fouling period, water permeability suddenly drops after the initial two days that this membrane is put in recirculation to be stabilized. Water permeability keeps decreasing slowly over time after a fast initial pressure drop and tends to

Figure 2.8 content:

Pressure Drop

—— Membrane A   —— Membrane B

Organic fouling   Biofouling

y-axis: dP* (bar) — 0.0, 0.1, 0.2, 0.3, 0.4, 0.5, 0.6, 0.7, 0.8

x-axis: Days — 0, 10, 20, 30, 40

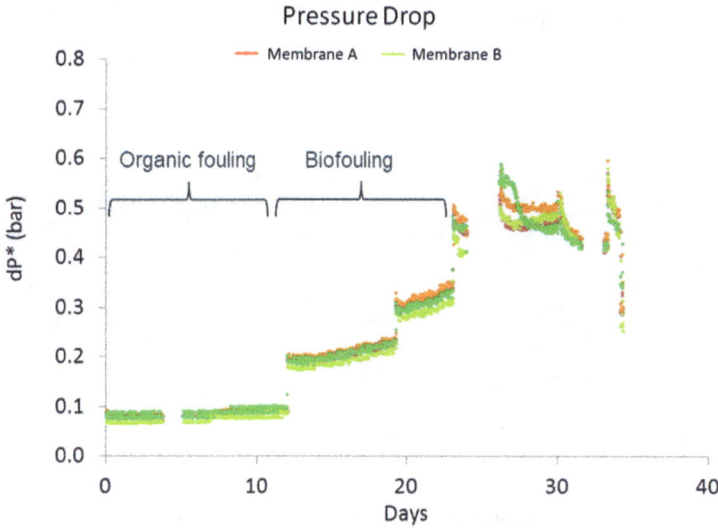

Figure 2.8: Pressure drop in an organic fouling followed by biofouling example.

plateau. This plateauing can be explained as the RO membrane works in cross-flow velocity; some foulants that are accumulated are removed as a result of the natural cross-flow operation. This sweeping of fouling, together with the accumulation of new fouling that reaches the membranes, establishes a fine balance that is able to explain this reduction of organic fouling accumulation over time. In this specific example, since biofouling is also developing over an already organically fouled membrane, it contributes to a further decrease of water permeability after the onset of the biological fouling period. The behavior of these fouling types on the membrane water permeability can be observed in Figure 2.9.

2.5.3 Salt passage

The salt passage across the membrane tends to follow a similar trend as the water permeability. This happens during the organic fouling period; as the membrane gets fouled, it is more difficult for water to pass through the membrane, but it is also more difficult for salt to pass through the membrane. This is why salt rejection can improve when a membrane is fouled. This behavior can be observed in Figure 2.10.

2.6 Quantifying the impact of biofouling

Biofouling has a severe impact on RO membrane systems. As bacteria start building a biofilm in the feed-concentrate channel of the membrane, the biofilm grows on the

A-value

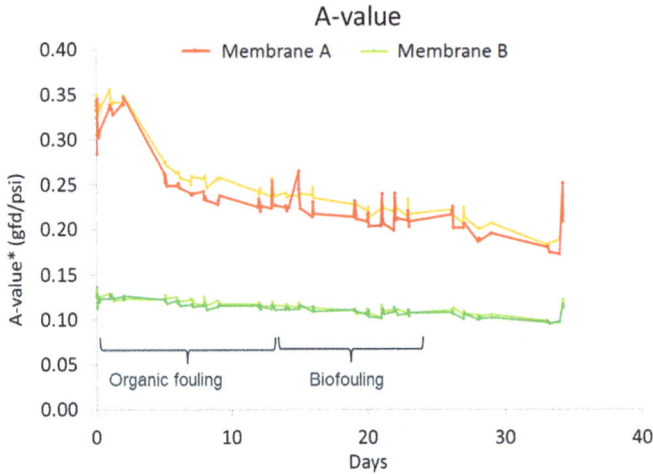

Figure 2.9: Water permeability (A-value) in an organic fouling followed by biofouling example.

B-value

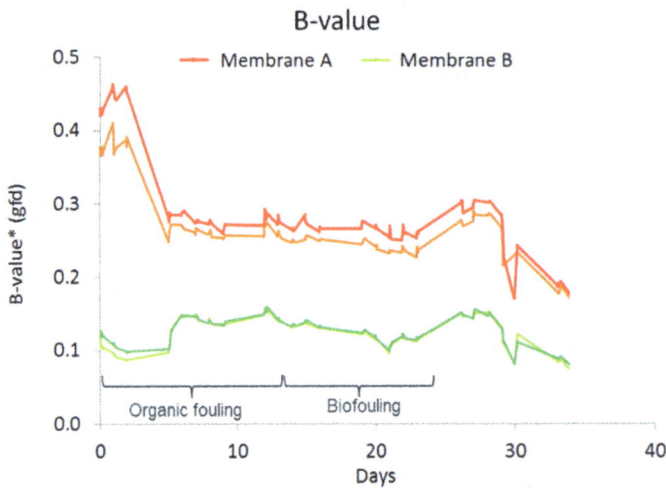

Figure 2.10: Example of salt permeance (B-value) in organic fouling followed by biofouling.

feed spacer as well as on the membrane surface. This causes an obstruction in the membrane, where water needs more energy to pass through it as the channel starts to get narrower as it gets clogged. In order to reverse this situation, a cleaning in place (CIP) can be done. Typically, a chemical cleaning can restore some of the original pressure drop value. Usually, the sooner the cleaning is performed, the more pressure drop that can be recovered. However, usually it is not possible to recover to the same pressure drop as that when the system started. And at some point, the membranes might need to be replaced. This replacement might come to reaching a point where the membrane cannot be operated under a reasonable pressure drop limit, or, as the membrane has been exposed to multiple chemical cleanings, it might start losing its performance over time. So, if an installation is designed to meet a certain water quality level on the permeate the exposure to multiple chemical cleanings might mean that the membrane cannot deliver the designed water quality any more and therefore it might need to be replaced. This leads to the stop of the RO installation, which means that now water is being produced during the membrane replacement period. This results in a cost increase in the total cost of water as well as affects the plant availability and productivity.

The impact of biofouling on RO membranes can be represented in terms of a reduction of the number of chemical cleanings for the same period of time. A typical period of time that is usually assumed is a yearly basis. An illustrative example on the impact that biofouling resistant membranes have in an RO system is presented in Figure 2.11. In this figure, it can be seen how after a defined operating time, and after setting up a certain pressure drop (dP) cleaning trigger, the conventional membrane reaches the cleaning trigger four times. If the time period is assumed to be one year, and the pressure drop cleaning trigger is assumed to be 3.5 bar, this will indicate that the RO system needs to be cleaned four times a year with a conventional membrane. On the other hand, thanks to the lower initial pressure drop that the fouling resistant membrane present, and thanks to the improved fouling resistance features of the membrane element, only three chemical cleanings are required for the same period of time and for the same pressure drop cleaning trigger. This represents a reduction of 25% on the number of chemical cleanings done on a yearly basis. The reduction of the number of chemical cleanings can also be expressed in terms of that the fouling resistant membrane will enable increasing the operating time before doing a chemical cleaning by 33%.

The cleanings reduction for a certain period of time can be calculated from the number of cleanings performed with a conventional membrane (Cleanings$_0$), and the number of cleanings performed with a fouling resistant membrane (Cleanings$_{fr}$). The cleanings reduction achieved thanks to a fouling resistant membrane can be calculated as follows:

$$\text{Cleanings reduction} = 1 - \frac{\text{Cleanings}_0}{\text{Cleanings}_{fr}} \qquad (2.1)$$

Using the example shown in Figure 2.11, and taking into account that the time scale is one year, it can be seen that the fouling resistant membrane will enable a reduction of 25% of the number of cleanings, effectively reducing the number of cleanings done in a year from 4 to 3.

Another way to quantify and express the reduction achieved thanks to fouling resistant membranes, or fouling mitigation technologies, is, instead of using the cleanings reduction, using the operating time increase. This value represents the additional time that the reverse osmosis membranes can be operated before having to stop the

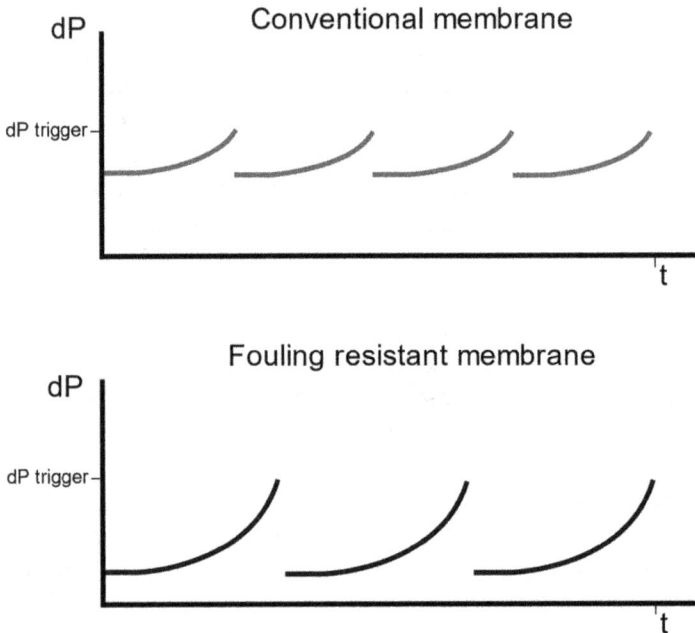

Figure 2.11: Diagram showing the impact of fouling resistant membrane.

whole membrane system in order to clean the membranes. The operating time increase can be calculated as follows:

$$\text{Operating time increase} = \frac{\text{Cleanings reduction}}{1 - \text{Cleanings reduction}} \tag{2.2}$$

Using the same example previously described, it can be calculated that for a reduction in the number of chemical cleanings of 25%, it would be possible to operate the RO membranes for 33% longer time without needing to stop in order to perform a chemical cleaning.

It is interesting to note the relationship that exists between the cleanings reduction and the operating time increase. This relationship can be observed in Figure 2.12. It can be seen that for a relatively small reduction in cleanings frequency, the increase in operating time is similar. As an example, a reduction of 20% in the number of chemical cleanings will equate to a 25% increase in operating time. However, a larger reduction in the chemical cleanings will result in an exponential increase in the operating time. As an example, a reduction of 70% of chemical cleanings will result in an increase of 233% in operating time.

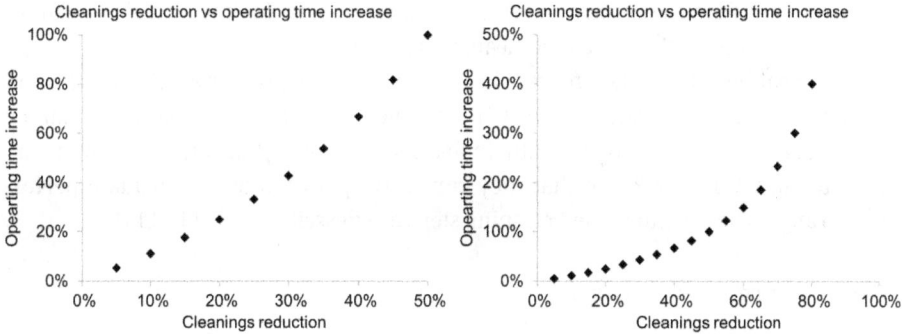

Figure 2.12: Relationship between the cleanings reduction and the operations time increase.

2.7 Cleaning fouled membranes

2.7.1 Cleaning biofouled membranes

Biofouling is generally the leading issue triggering cleaning in industrial wastewater treatment plants. Although cleaning guidelines recommend performing a CIP when the normalized pressure drop increases by 10–15%, it is observed that some plants clean at the maximum allowed vessel pressure drop of 3.5 bar [13]. This maximum limit is standard for 8-inch RO elements in order to avoid irreversible mechanical damage to the elements. When dealing with biofouling, it is recommended to clean as soon as possible, as it is observed that when membranes are not cleaned earlier, and pressure drop starts to increase; the more it increases, the less effective are the chemical cleanings in restoring the initial pressure drop performance, and therefore, the biofouling degree of the membrane system accelerates and becomes worst.

It should be noticed that in order to assess the normalized pressure drop increase of an RO system, the operating data of the RO installation needs to be normalized. Data normalization is described in this book, and it is a great tool to understand the whole situation of an RO system, and to assess if the membranes are suffering from fouling or they are operating normally.

The chemical cleaning typically preferred to perform the CIP consists of a caustic solution that uses caustic soda (NaOH), at pH 13, at a maximum temperature of 35 °C. It should be noted that biological fouling is more challenging to clean than organic fouling. This is why compared to organic fouling, the recommended pH and temperate are higher, as it has been shown that a high pH, coupled with a high temperature is the most effective combination for cleaning biologically fouled membranes. It should also be taken into consideration that typically biofouling comes coupled with a membrane that has already been suffering from organic fouling. However, this cleaning strategy, as it is harsher than the recommended cleaning for organic fouling, will also clean organic fouling while at the same time cleaning biofouling.

This cleaning can be then followed by an acid cleaning that uses HCl, at pH 2 and with a maximum temperature of 45 °C, to remove any inorganic material that can be present together with the biologically fouled membrane and also to allow the membrane to condition the RO so that they can start operation at an adequate performance range. The recommended cleaning steps are described in Table 2.1 [13].

Table 2.1: Cleaning sequence for a membrane suffering from biofouling.

Step	Description
1	Cleaning with an NaOH chemical solution at pH 13 with a maximum temperature of 35 °C.
2	Recirculate the solution for 30 min. It should be noted that if the color in the chemical solution changes, the chemical solution should be disposed and Step 1 should be repeated.
3	Leave the membranes with a soaking using the prepared chemical solution.
4	Perform a high-flow recirculation using the prepared chemical solution.
5	Flush out the chemical solution.
6	Repeat Steps 1 to 5 with a HCl chemical solution at pH 2 with a maximum temperature of 45 °C.

2.7.2 Cleaning organic fouled membranes

When organic fouling happens in an RO system, a CIP is typically the recommended option to attempt to restore the initial water permeability to the system. An early sign that the membrane might be suffering from organic fouling is when the normalized permeate flow increases by 10% [13]. In order to assess if the normalized permeate flow has increased over time, the plant operating data should be normalized. The steps to normalize the operating data that allows understanding the performance of the RO installation is described in this book.

The chemical cleaning typically preferred to perform the CIP consists of a caustic solution that uses caustic soda (NaOH), at a pH 12, at a maximum temperature of 30 °C. This cleaning can be then followed by an acid cleaning that uses HCl, at pH 2 and with a maximum temperature of 45 °C, to remove any inorganic material that can be present together with the organically fouled membrane and also to allow the membrane to condition the reverse osmosis membranes so that they can start its operation at an adequate performance range. The recommended cleaning steps are described in Table 2.2 [13].

Table 2.2: Cleaning sequence for a membrane suffering from organic fouling.

Step	Description
1	Cleaning with an NaOH chemical solution at pH 12 with a maximum temperature of 30 °C.
2	Recirculate the solution for 30 min. It should be noted that if the color in the chemical solution changes, the chemical solution should be disposed and Step 1 should be repeated.
3	Leave the membranes with a soaking using the prepared chemical solution.
4	Perform a high-flow recirculation using the prepared chemical solution.
5	Flush out the chemical solution.
6	Repeat Steps 1 to 5 with a HCl chemical solution at pH 2 with a maximum temperature of 45 °C.

2.8 Designing biofouling-resistant membranes

RO membranes that are resistant to biofouling exist in the market, and they present certain features. The main characteristics that these membranes possess are usually characterized by three factors.

The first one involves having a biofouling resistant spacer, which usually is designed to minimize the dead spaces where bacteria can get sheltered, and comfortably receive nutrients so that they can grow and eventually build a biofilm. These are typically zones where the cross-flow velocities are low or close to zero. Additionally, feed spacers that have an initial pressure drop also contribute to a decrease in the biofouling rate, as usually biofouling that would develop on these membranes need more time to reach the pressure drop cleaning trigger. This can lead to a decrease in the chemical cleanings rate, which also leads to an increase in the operating time between cleanings [34].

The second one involves designing a membrane that has biofouling resistant properties. This makes it more difficult for bacteria to attach and start growing a biofilm on the membrane surface. This might lead to a difficulty in bacterial growth and therefore to develop, as overall, the environment is less suitable for a biofilm to develop and biologically foul the reverse osmosis membrane [35, 36].

The third strategy in designing a biofouling resistant membrane involves improving the overall reverse osmosis membrane element design and construction, so that dead zones, also called stagnant zones, where bacteria can attach are minimized. This means that bacteria cannot find a suitable space to settle and quietly grow without being disturbed. Bacteria need to find these places where cross-flow velocity is minimized so they can slowly grow over time and receive a constant supply of nutrients [37].

References

[1]   C. Fritzmann, J. Löwenberg, T. Wintgens, T. Melin. State-of-the-art of reverse osmosis desalination. Desalination, 216(1) (2007) 1–76.

[2]   R. Valavala, J. Sohn, J. Han, N. Her, Y. Yoon. Pretreatment in reverse osmosis seawater desalination: A short review. Environmental Engineering Research, 16(4) (2011) 205–212.

[3]   L. F. Greenlee, D. F. Lawler, B. D. Freeman, B. Marrot, B. P. Moulin. Reverse osmosis desalination: Water sources, technology, and today's challenges. Water Research, 43(9) (2009) 2317–2348.

[4]   A. Matin, Z. Khan, S. M. J. Zaidi, M. C. Boyce. Biofouling in reverse osmosis membranes for seawater desalination: Phenomena and prevention. Desalination, 281 (2011) 1–16.

[5]   M. G. Khedr. Membrane fouling problems in reverse-osmosis desalination applications. International Desalination and Water Reuse Quarterly, 10(3) (2000) 8–17.

[6]   D. H. Paul. Reverse osmosis: Scaling, fouling and chemical attack. Desalination Water Reuse, 1 (1991) 8–11.

[7]   S. P. Chesters, N. Pena, S. Gallego, M. Fazel, M. W. Armstrong, F. Del Vigo. Results from 99 seawater RO membrane autopsies. IDA Journal of Desalination and Water Reuse, 5(1) (2013) 40–47.

[8]   DuPont. Interviewing customers on the importance of biofouling. DuPont Water Solutions Internal Report, (2019).

[9]   W. Guo, H. H. Ngo, J. Li. A mini-review on membrane fouling. Bioresource Technology, 122 (2012) 27–34.

[10] T. Carroll, S. King, S. R. Gray, B. A. Bolto, N. A. Booker. The fouling of microfiltration membranes by NOM after coagulation treatment. Water Research, 34(11) (2000) 2861–2868.

[11] L. D. Nghiem, P. J. Coleman, C. Espendiller. Mechanisms underlying the effects of membrane fouling on the nanofiltration of trace organic contaminants. Desalination, 250(2(2010) (2010) 682–687.

[12] K. O. Agenson, T. Urase. Change in membrane performance due to organic fouling in nanofiltration (NF)/reverse osmosis (RO) applications. Separation and Purification Technology, 55(2) (2007) 147–156.

[13] DuPont™. DuPont Water Solutions FilmTec™ Reverse Osmosis Membranes Technical Manual. Form No. 609-00071-1009.

[14] M. Herzberg, K. Seoktae, M. Elimelech. Role of extracellular polymeric substances (EPS) in biofouling of reverse osmosis membranes. Environmental Science & Technology, 43(12) (2009) 4393–4398.

[15] A. Matina, S. M. J. Khana, M. C. Zaidiam, M. C. Boyceb. Biofouling in reverse osmosis membranes for seawater desalination: Phenomena and prevention. Desalination, 281 (2011) 1–16.

[16] M. Herzberg, M. Elimelech. Biofouling of reverse osmosis membranes: Role of biofilm-enhanced osmotic pressure. Journal of Membrane Science and Technology, 295(1–2) (2007) 11–20.

[17] T. H. Chong, F. S. Wong, A. G. Fane. The effect of imposed flux on biofouling in reverse osmosis: Role of concentration polarisation and biofilm enhanced osmotic pressure phenomena. Journal of Membrane Science and Technology, 325(2) (2008) 840–850.

[18] M. Herzberg, M. Elimelech. Biofouling of reverse osmosis membranes: Role of biofilm-enhanced osmotic pressure. Journal of Membrane Science, 295(1–2) 2007).11–20.

[19] J. S. Baker, L. Y. Dudley. Biofouling in membrane systems – A review. Desalination, 118(1–3) (1998) 81–89.

[20] T. Knoell, J. Safarik, T. Cormack, R. Riley, S. W. Lin, H. Ridgway. Biofouling potentials of microporous polysulfone membranes containing a sulfonated polyether-ethersulfone/polyethersulfone block copolymer: Correlation of membrane surface properties with bacterial attachment. Journal of Membrane Science, 157(1) (1999) 117–138.

[21] O. Sánchez. Microbial diversity in biofilms from reverse osmosis membranes: A short review. Journal of Membrane Science, 545 (2018) 240–249.

[22] C. Dreszer, J. S. Vrouwenvelder, A. H. Paulitsch-Fuchs, A. Zwijnenburg, J. C. Kruithof, H. C. Flemming. Hydraulic resistance of biofilms. Journal of Membrane Science, 429 (2013) 436–447.

[23] J. Wingender, H. C. Flemming. Contamination potential of drinking water distribution network biofilms. Water Science and Technology, 49(11–12) (2004) 277–286.

[24] U. Bertheas, K. Majamaa, A. Arzu, R. Pahnke. Use of DBNPA to control biofouling in RO systems. Desalination & Water Treatment, 3(1–3) (2009) 175–178.

[25] H. J. Lee, M. A. Halali, S. Sarathy, C. F. De Lannoy. The impact of monochloramines and dichloramines on reverse osmosis membranes in wastewater potable reuse process trains: A pilot-scale study. Environmental Science: Water Research & Technology, 6(5) (2020) 1336–1346.

[26] P. Trayhurn. Oxygen – A critical, but overlooked, nutrient. Frontiers in Nutrition, 6 (2019) 10.

[27] J. D. Jacobson, M. D. Kennedy, G. Amy, J. C. Schippers. Phosphate limitation in reverse osmosis: An option to control biofouling?. Desalination & Water Treatment, 5(1–3) (2009) 198–206.

[28] J. S. Vrouwenvelder, F. Beyer, K. Dahmani, N. Hasan, G. Galjaard, J. C. Kruithof, M. C. M. Van Loosdrecht. Phosphate limitation to control biofouling. Water Research, 44(11) (2010) 3454–3466.

[29] G. Massons, G. Gilabert-Oriol, S. Arenas-Urrera, J. Pordomingo, J. C. González-Bauzá, E. Gasia, M. Slagt. Industrial scale pilot at Maspalomas I desalination plant demonstrates the efficiency of DuPont™ B-Free™ pretreatment–a new breakthrough solution against biofouling. Desalination & Water Treatment, 259 (2022) 261–265.

[30] G. Massons, G. Gilabert-Oriol, M. Slagt, R. Balakrishnan, H. Pandya, A. Elsayed, H. Alomar. Testing of DuPont™ B-Free™ technology in Arabic Gulf water at Sharjah Electricity & Water Authority (SEWA) Hamriyah Desalination Plant. Desalination and Water Treatment, 309 (2023) 80–83.

[31] N. M. Farhat, J. S. Vrouwenvelder, M. C. Van Loosdrecht, S. S. Bucs, M. Staal. Effect of water temperature on biofouling development in reverse osmosis membrane systems. Water Research, 103 (2016) 149–159.

[32] G. Gilabert-Oriol, L. Hailemariam, G. Massons, W. Su. System for determining filter status, World Intellectual Property Organization. (WIPO) Patent Application WO/2024/238283, (2024)

[33] G. Gilabert-Oriol, L. Hailemariam, G. Massons, W. Su, Z. Liao, Z. Jensen. System for determining filter status. World Intellectual Property Organization (WIPO) Patent Application WO/2024/235116, (2024)

[34] H. Maddah, A. Chogle. Biofouling in reverse osmosis: Phenomena, monitoring, controlling and remediation. Applied Water Science, 7 (2017) 2637–2651.

[35] H. Karkhanechi, R. Takagi, H. Matsuyama. Biofouling resistance of reverse osmosis membrane modified with polydopamine. Desalination, 336 (2014) 87–96.

[36] P. S. Goh, A. K. Zulhairun, A. F. Ismail, N. Hilal. Contemporary antibiofouling modifications of reverse osmosis desalination membrane: A review. Desalination, 468 (2019) 114072.

[37] E. M. Hoek, T. M. Weigand, A. Edalat. Reverse osmosis membrane biofouling: Causes, consequences and countermeasures. NPJ Clean Water, 5(1) (2022) 45.

Chapter 3
Membrane fouling simulators: a quick tool for studying biofouling and why biofouling does not depend on flux

Comparing biofouling development in membrane fouling simulators and spiral-wound reverse osmosis elements using river water and municipal wastewater

Membrane fouling simulators (MFSs) are flat cell units used to simulate the biofouling development of spiral-wound reverse osmosis (RO) elements. In this research, MFS units and two RO testing systems were operated in parallel, using different water types on each test. Differences on differential pressure increase and fouling distribution between the two pilot plants were evaluated. In the first study, several RO elements and MFS units were operated with the same conditions to assess the reliability of the testing systems. In the second study, the performance of different feed spacer types assembled in full-scale RO elements was compared to the same feed spacer types assembled in the MFS unit. The results of these studies showed that the relative biofouling impact in the MFS units was equivalent to the performance of the RO elements. Additionally, the results from the second study provide indications that a prototype 28-mil feed spacer (28 T1) may provide significantly more biofouling resistance than any of the 34-mil feed spacers evaluated in this study.

3.1 Introduction

Water scarcity is recognized as one of the main threats that mankind is facing globally [1]. RO membrane technology has developed as a promising, cost-effective technology to remove contaminants from non-potable waters and provide fresh water supply to meet the growing demand [2]. RO elements, however, can suffer from progressive loss of performance when treating challenging waters due to fouling [3]. Of all fouling

Acknowledgments: The author would like to acknowledge the contributors of this chapter: Guillem Gilabert-Oriol, Gerard Massons, Diana Dubert, Jon Johnson, and Tina Arrowood. The author would also like to thank all DuPont Water Solutions team for their support on doing this research, and specifically Ana Antolín for her great effort in setting up the MFS assets, as well as Nicolas Corgnet and Javier Dewisme for their outstanding help in performing the experiments.

This chapter was originally published with the following reference: G. Massons-Gassol, G. Gilabert-Oriol, J. Johnson, T. Arrowood, Comparing biofouling development in membrane fouling simulators and spiral-wound reverse osmosis elements using river water and municipal wastewater. Industrial & Engineering Chemistry Research, 56(40) (2017) 11628–11633.

https://doi.org/10.1515/9783111639079-003

types, biofouling is one of the most complex to manage in RO water treatment systems [4]. It occurs when bacteria colonize and form biofilms in the feed channel of the RO elements, causing increased friction for water flow. This causes the differential pressure (dP) to increase [5], leading to hydraulic imbalance and, if not controlled, can damage the element. Additionally, biofilms can affect membrane transport properties, as the polymeric film formed on the membrane surface decreases the overall water permeability [6]. Each of these effects increase the energy of operation and leads to frequent system shutdowns for chemical cleanings to regain membrane performance. The high pH conditions needed to remove biofilms during cleaning can result in membrane hydrolysis and shorten the useful life of the element. Therefore, in total, system productivity, chemical usage, membrane life, and energy, each contribute to higher cost of water production when biofouling occurs. Mechanisms to control biofouling are needed to enable long term performance when treating water with high contamination levels [7].

Studying biofouling in water treatment systems is complex due to the multiple variables that affect biofilm formation [5]. To accelerate research, screening tools that enable biofouling experiments to be conducted with different water types and enable multiple parameters to be explored in parallel are needed.

MFSs have been described as a cost-effective tool to predict biofouling evolution in full-scale RO systems [8]. Differential pressure in the MFS models the increase of an RO system, since the biofilm on an RO system generally starts in the first centimeters of the feed-concentrate channel [9–10]. Thus, MFS units can be used to study biofouling and quickly screen new solutions without having to manufacture an entire RO element [11].

MFS units are especially suited for testing one of the key parameters influencing biofouling in spiral-wound RO elements, the feed spacer [12]. The main role of the feed spacer is to promote turbulence and improve mass transfer by distorting the laminar profile of the axial flow when operated in cross-flow [13]. However, low shear stress zones from flow stagnation are created by the feed spacer [14], and simulations suggest that these are the areas where biofilm develops more strongly [15]. Defining feed spacer design features (e.g., thickness, strand angle, and spacing between filaments), to improve biofouling resistance and reduce the rate of element pressure drop increase has been a topic of recent interest [16–18].

The aim of this work is to demonstrate the ability of MFS units to model biofouling that occurs in RO elements when treating water with high fouling potential. Additionally, the impact of feed spacer type on the rate of differential pressure increase due to biofouling is explored in a set of RO elements and MFS units. If results obtained are comparable, these will demonstrate the usefulness of the MFS units as a screening tool for expected element performance under biofouling environments.

3.2 Materials and methods

3.2.1 Membrane fouling simulators

The MFS units are portable and can be installed in a feed water side-stream as a stand-alone test unit or in parallel with full-scale RO systems. A picture and a diagram of the MFS are shown in Figure 3.1. MFS are small units that simulate the feed channel of an RO element by layering feed spacer on top of membrane fitted in a rectangular flow cell. Water is directed through the feed spacer channel at a set flow rate but does not permeate. The differential pressure across the MFS feed channel is monitored during the experiment. In the present study, four MFS units (MFS1, MFS2, MFS3, and MFS4) were operated in parallel. The transparent cells were assembled with a feed spacer and membrane coupon (20-cm length and 4-cm width) in each. Manual readings from the pressure drop indicator were recorded twice per day during the course of each trial. The pressure drop as a function of time from the MFS units was compared to the respective RO element systems that ran in parallel.

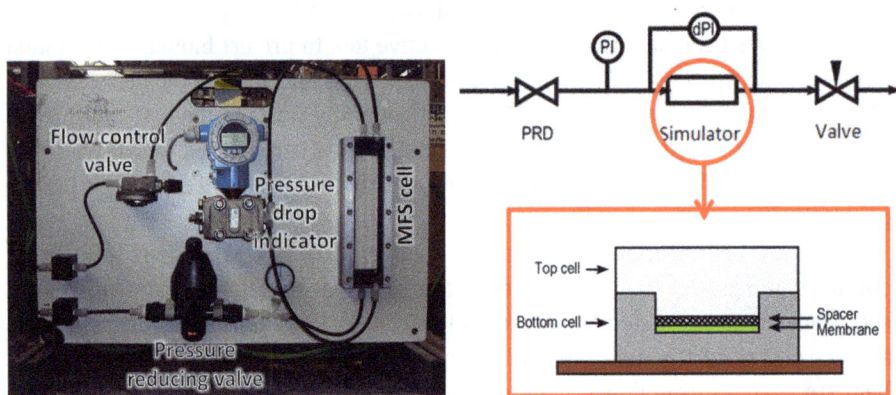

Figure 3.1: MFS simulator picture and diagram.

3.2.2 RO element testing units

3.2.2.1 The 2.5-inch multielement RO test

The pilot plant is configured with eight pressure vessels operated in parallel, each holding one 2.5-inch membrane elements by 14-inch-long spiral-wound RO membrane element. The membrane selected is therefore the 2514 membrane element. A single feed pump provides the feed to all eight pressure vessels. This system has a sampling valve after the pump to use the side-stream as feed water to operate the MFS units in parallel. The testing unit diagram can be found in Figure 3.2. This experiment was performed in the Global Water Technology Center that DuPont Water Solutions has in Tarragona, Catalonia, Spain.

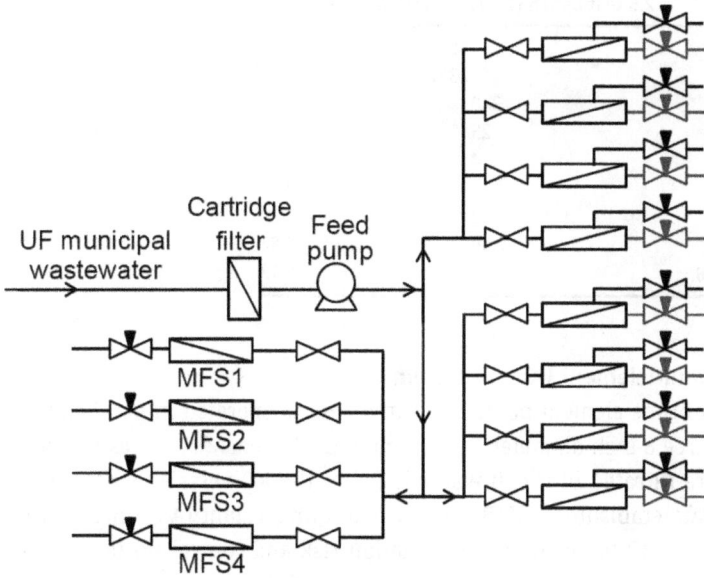

Figure 3.2: Configuration of the 2.5-inch reverse osmosis testing unit with MFS installed in parallel.

The 2514 membrane elements used in each vessel had identical element designs (0.55 m^2 membrane area and 22-mil feed spacer) with the aim to provide seven replicate data points and measure vessel-to-vessel reproducibility. The elements were controlled and operated at the same feed flow and recovery and the feed-to-concentrate press drop of each vessel was monitored over the course of the experiment. Specific operating conditions for each 2.5-inch pressure vessel (PV) and MFS are provided in Table 3.1. Four MFS units were installed in parallel to the 2514 elements. Each transparent MFS cell was assembled with the same membrane and feed spacers type (34-mil feed spacer thickness) in order to provide four replicates. The cross-flow velocity for each system was calculated [19] based on spacer thickness, dimensions of the 2514 element or MFS cell, and feed flow.

At the end of the fouling period, the 2514 elements were removed from the pressure vessels and autopsied, and the MFS cells were disassembled. Samples of feed spacer and membrane from each pilot plant were used to quantify and compare the adenosine triphosphate (ATP) and total organic carbon (TOC) accumulated during the experiment.

Table 3.1: The 2.5 test bench and MFS unit conditions used during the wastewater test.

Parameter	2.5 test bench (7 PV)	MFS units (4 units)
Duration (days)	46	
Temperature (°C)	23–27	
Feed flow (L/h)	340	16
Flux (L/m² h)	25	–
Recovery (%)	4	–
Membrane area (m²)	0.55	0.008
Spacer type (mil)	22	34
Spacer thickness (mm)	0.56	0.86
Cross-flow velocity (m/s)	0.09	0.13

3.2.2.2 Eight-inch multielement RO test system

The 8-inch RO membrane element bench has three vessels operated in parallel. Each is equipped with three 8-inch diameter by 40-inch-long elements (8,040), as shown in Figure 3.3. The ultra-filtrated brackish water is fed to the RO vessels using a single feed pump. Sodium metabisulfite (SMBS, 3 mg/L) and antiscalant (AS, 1 mg/L) were dosed upstream of the RO to avoid any chlorination risk and control scaling, respectively. The four MFS units were connected upstream of the feed pump but downstream of the chemical injection point.

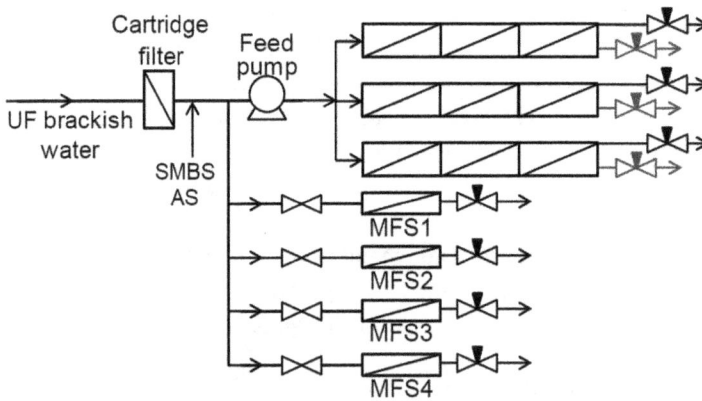

Figure 3.3: An 8-inch reverse osmosis testing unit with MFS simulators in parallel.

Each pressure vessel and MFS unit was equipped with different feed spacer types as summarized in Table 3.2. The performance of the different feed spacer designs was tested under biofouling conditions. Three of the feed spacers were tested in both 8-inch membrane elements and MFS cells, in order to confirm the reliability of the pressure drop results.

The two systems were adjusted so that the inlet cross-flow velocity for each feed spacer configuration was the same between the 8-inch test bench and the MFS unit. Despite the feed water temperature being only 15–18 °C, biofouling developed very quickly, and pressure drop increased after only one week of operation. Upon completion of the testing, only membrane and spacer samples from the MFS units were analyzed for biofilm quantification. The 8-inch elements were allowed to continue to operate for long term performance testing.

Table 3.2: Operating conditions 8-inch element versus MFS comparison.

Parameter	8-Inch element test bench			MFS unit			
	PV1	PV2	PV3	MFS1	MFS2	MFS3	MFS4
Spacer type	28 STD	34 STD	34 T1	28 STD	34 STD	28 T1	34 T1
Spacer thickness (mm)	0.71	0.86	0.86	0.71	0.86	0.71	0.86
Membrane area (m²)	111.4	111.4	111.4	0.008	0.008	0.008	0.008
Feed flow (L/h)	8,400	8,400	8,400	16	16	16	16
Inlet cross-flow velocity (m/s)	0.16	0.13	0.13	0.16	0.13	0.16	0.13
Flux (L/m²h)	24	24	24		–		
Recovery (%)	33	33	33		–		

3.2.3 Analytical biofouling characterization

After completing each trial, the fouling distribution on the MFS was examined using a Leica MS5 stereoscope. Additionally, membrane and feed spacer samples were taken from the MFS coupons and the RO elements. Fouling was extracted using the protocol described in previous studies [19]. The extracted foulant was analyzed for TOC using a Shimadzu TOC-L analyzer and ATP using a Celsis Advance Luminometer.

3.2.4 Water sources

Water from the secondary effluent of the Vilaseca wastewater treatment plant and the Ebro River provide the feed water to the RO with 2.5-inch membrane elements and the 8-inch membrane element systems, respectively. Both wastewater and river water are pretreated with DuPont™ ultrafiltration modules prior to feeding to the downstream the RO units. Table 3.3 summarizes the basic water composition of each water source at the RO feed point. Both waters are of the brackish type and have the essential ingredients for biofouling, TOC, ATP, nitrate, and phosphate.

Table 3.3: Vilaseca wastewater and Ebro River water characterization.

Parameter	Vilaseca wastewater (2.5-inch RO element test)	Ebro River water (8-inch RO element test)
TDS (mg/L)	1,440	1,000
TOC (mg/L)	6.6	1.3
ATP (ng/L)	83	6
Nitrate (mg/L)	30	11
Phosphate (mg/L)	0.55	0.04

3.3 Results and discussion

3.3.1 MFS versus 2.5-inch element test bench

The MFS units were expected to show a similar biofouling characteristic as the elements, despite having less surface area to harbor bacteria. The ratio of membrane length between an MFS cell (4 cm × 20 cm) and a 2514 membrane element (6.4 cm diameter × 35.5 cm long) is 1:1.8. Using municipal wastewater as feed, the rate of biofouling growth on the 2.5-inch membrane elements and MFS cells was compared. Additionally, the reproducibility of each pilot unit was evaluated by comparing the average dP of seven elements and four MFS cells. As described previously, all the spiral-wound elements were constructed and operated at the same conditions. Similarly, the MFS units each were assembled with the same membrane and feed spacer and operated at the same feed flow.

The biofouling rate of the Vilaseca wastewater was moderate. It took nearly 2 weeks to see a measurable increase in the differential pressure. These pressure drop increase evolution over time can be found in Figure 3.4. The measurement variability between the 2.5-inch elements was about 8%, while the MFS units have slightly higher variability (17%) due to manual operation and smaller sample size. However, they provided excellent reproducibility of the stages of biofilm development on the spiral-wound elements.

The agreement between the 2.5-inch membrane elements and the MFS cells performance is illustrated in Figure 3.5, where the daily average change in differential pressure is compared between the two units. The high correlation observed between the two sets of results was proved to be statistically significant using JMP® Pro 12.2.0 with a calculated p-value lower than 0.0001. However, variability increases at higher degrees of biofouling. Thus, when comparing different element types or MFS assemblies, like comparing different feed spacers or membrane types, it looks more reliable to compare performance in the early stages of biofilm growth.

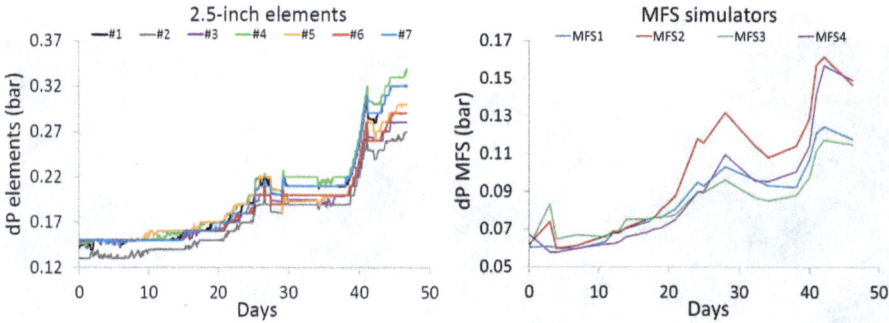

Figure 3.4: Pressure drop of MFS and 2.5-inch elements.

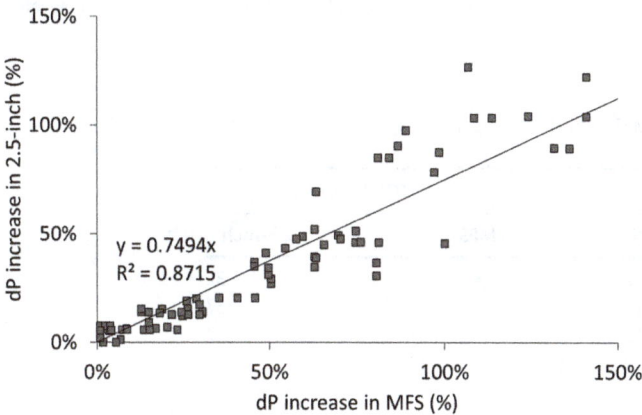

Figure 3.5: Correlation in differential pressure increase of MFS and 2.5-inch elements.

The biofilm formed in the MFS unit was visually more concentrated on the feed spacer strands than the membrane surface. This biofouling distribution in the inlet and outlet of the membrane and feed spacer is shown in Figure 3.6. This is similar to the observation of the biofilm in the autopsied element, where the inlet part was showing more biofouling accumulation than the outlet area. The biofilm was a sticky brownish slime mostly attached on the feed spacer strands.

The ATP and TOC analysis confirmed the presence of biofouling in both units, as summarized in Table 3.4, and provides a useful measurement to compare the extent of biofouling in each system. Among the replicates, the standard deviation of the measurements was about 14%, which was in agreement with similar analyses found in the literature [20, 21].

Inlet   Outlet

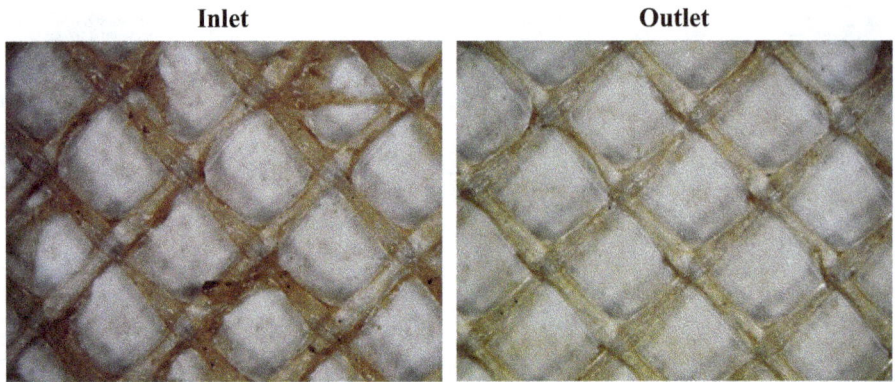

Figure 3.6: Magnified image of the membrane and spacer assembly in the MFS after operation (flow left to right).

Table 3.4: ATP evaluation in MFS simulators compared to spiral-wound elements.

ATP (ng/cm^2)		TOC (mg/m^2)	
MFS	2.5-Inch	MFS	2.5-Inch
8.7	7.9	19	32
8.3	13	18	37
8.9	12	21	47
5.8	8.2	20	38
–	9.2	–	45
–	11	–	49
–	11	–	34
Av.: 8 ± 1	Av.: 10 ± 2	Av.: 19 ± 1	Av.: 40 ± 6

3.3.2 MFS versus 8-inch multielement system

When comparing the MFS units to a three-element 8-inch pressure vessel (20.3 cm diameter × 303 cm long), the ratio of total length, 1:15.2, is much greater than for the 2.5-inch membrane single element, 1:1.8. The ability for the MFS to provide comparative relative pressure drop performance to full-scale system was examined. In this study, each pressure vessel was equipped with three 400 ft^2 active area elements containing different feed spacer types. Three of the MFS units were assembled with the same three types of spacers used in the elements and the fourth MFS unit used a prototype 28-mil feed spacer. The feed spacers used in this study can be characterized based on the angle, spacing between strands, and thickness, as represented in Figure 3.7.

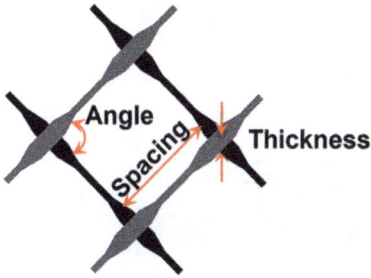

Figure 3.7: Feed spacer geometry characterization.

Details of the different feed spacers use in each system are summarized in Table 3.5. Due to the proprietary nature of the prototype, T1, feed spacers directional arrows are used to show their relative measurement as compared to the standard version of the same thickness.

Table 3.5: Details of the different feed spacers assessed.

Spacer	Thickness	Spacing	Angle	MFS	8-Inch
28 STD	0.71 mm	2.82 mm	90°	✓	✓
34 STD	0.86 mm	2.82 mm	90°	✓	✓
28 T1	0.71 mm	↑	↓	✓	–
34 T1	0.86 mm	↑	↓	✓	✓

The start of this test began with the four parallel MFS units tied in to the 8-inch membrane element system feed line. The MFS unit contained the same membrane type and feed spacer designs as the 8-inch elements. Care was taken to ensure the MFS units were operated with the same cross-flow velocity as the RO module, as described in Table 3.2. Doing so provided the most comparable biofouling environment for evaluating the performance of the MFS unit relative to the 8-inch element. It also allowed a comparison of the feed spacer hydrodynamics, by comparing their initial pressure drop. Figure 3.8 summarizes the differential pressure data collected during the experiment in both plants. The relative order of differential pressure measured in the 8-inch system was the same as the MFS units, from higher to lower: 28 STD, 34 STD, and 34 T1. The absolute differential pressure registered on the MFS for the 34 STD configuration was slightly lower than on the spiral-wound element. This could be caused by the inherent variability due to the manual assembly of the MFS units described in Section 3.3.1. Also noteworthy in the MFS comparison, the differential pressure of the 28 T1 feed spacer was significantly lower than the two 34-mil feed spacers. This data shows that in addition to feed spacer thickness, spacer geometry can impact the initial pressure drop and pressure drop increase.

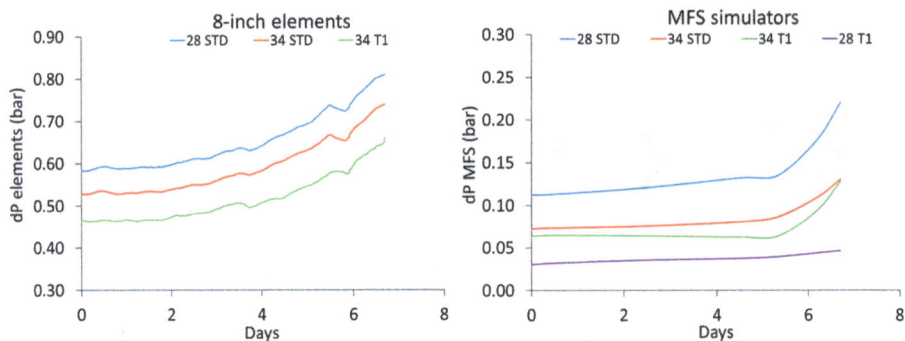

Figure 3.8: Differential pressure rises in 8-inch element test bench and MFS units.

The Ebro River water provided very quick biofouling in both the 8-inch and MFS systems as indicated by the rapid pressure drop increase within 7 days of operation, as can be observed in Figure 3.8. The performance of each spacer type gave similar relative differential pressure performance as the biofilm developed and the pressure drop increased. The initial increase in pressure drop in the 8-inch element system was not captured in the MFS units because the pressure readings were manual, and the rise occurred over the weekend. However, in the last phase, the trends were very similar in both plants.

After operation, the MFS were disassembled and the foulant was extracted from the membrane and feed spacer sample. ATP and TOC concentrations were measured and are summarized in Table 3.6. Significant TOC accumulated in each unit and the ATP levels confirm that there was biological activity, in accordance with the operational results. The sample taken from the 28 T1 unit showed the lowest amount of fouling among all tested spacers. This agrees with the observed low differential pressure rise during the testing. Based on these promising results a more thorough investigation to confirm the improved biofouling performance of the 28 T1 spacer will be explored in the future.

Table 3.6: Biofouling quantification in MFS coupons.

Feed spacer type	ATP (ng/cm^2)	TOC (mg/m^2)
28 STD	2.7	98
34 STD	3.0	56
34 T1	5.2	103
28 T1	2.9	33

The biofouling distribution over the coupons was visually examined. In each unit, more biofouling was accumulated on the inlet side of the MFS than on the outlet side.

Figure 3.9 shows the results from the 34-mil STD coupon. This observation is in agreement with other studies that show that most of the biofouling is accumulated on the feed side of the first element of an RO system [22]. The distribution of biofouling accumulation was similar to that observed on the fouled MFS coupons from the wastewater test, as can be visualized in Figure 3.6. The biofilm was a sticky yellowish slime mostly attached on the feed spacer strands.

Inlet   **Outlet**

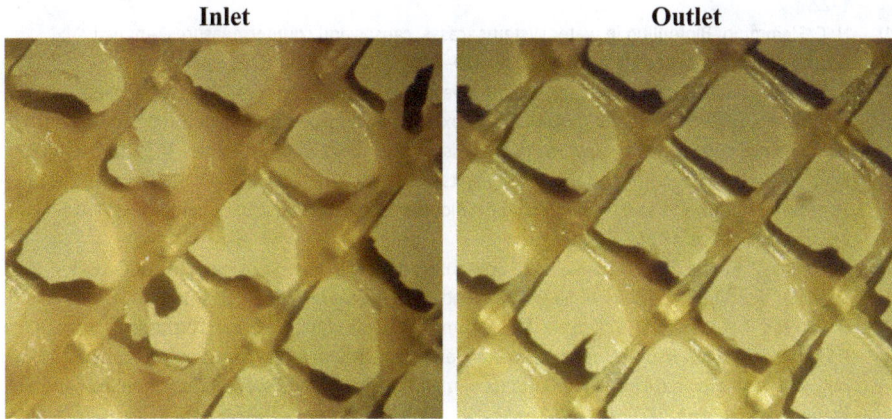

Figure 3.9: Visual comparison of the inlet and outlet end of the 34 STD coupon (flow left to right).

3.4 Conclusions

The MFS units showed a similar biofouling as spiral-wound RO systems, when exposed to river water and municipal wastewater. The MFS simulators can be useful screening tools to evaluate the pressure drop performance of large spiral-wound elements as well as different module designs, such as feed spacer, during biofouling events.

The magnitude of the differential pressure changes were not the same in the MFS and element systems due to the difference in scale. However, similar relative trends were observed and demonstrated the usefulness of the MFS units as a screening tool for biofouling in RO units.

This study also showed promising biofouling resistance of a 28 T1 spacer with both lower relative pressure drop increase during operation and lower biofouling accumulation, compared to three other spacer designs. These qualities could enable saving in energy required to operate the RO elements and extended performance when treating challenging water. Further examination of the biofouling resistance of this spacer type will be the subject of future studies.

References

[1]   C. Fritzmann, J. Löwenberg, T. Wintgens, T. Melin. State-of-the-art of reverse osmosis desalination. Desalination, 216 (2007) 1–76.

[2]   L. F. Greenlee, D. F. Lawler, B. D. Freeman, B. Marrot, P. Moulin. Reverse osmosis desalination: Water sources, technology, and today's challenges. Water Research, 43 (2009) 2317–2348.

[3]   K. P. Lee, T. C. Arnot, D. Mattia. A review of reverse osmosis membrane materials for desalination – development to date and future potential. Journal of Membrane Science and Technology, 370 (2011) 1–22.

[4]   H. C. Flemming. Biofouling in water systems–cases, causes and countermeasures. Applied Microbiology and Biotechnology, 59 (2002) 629–640.

[5]   C. Sun, L. Fiksdal, A. Hanssen-Bauer, M. B. T. Rye, T. Leiknes. Characterization of membrane biofouling at different operating conditions (flux) in drinking water treatment using confocal laser scanning microscopy (CLSM) and image analysis. Journal of Membrane Science, 382 (2011) 194–201.

[6]   S. R. Suwarno, X. Chen, T. H. Chong, V. L. Puspitasari, D. McDougald, Y. Cohen, S. A. Rice, A. G. Fane. The impact of flux and spacers on biofilm development on reverse osmosis membranes. Journal of Membrane Science, 405 (2012) 219–232.

[7]   A. Matina, Z. Khana, S. M. J. Zaidia, M. C. Boyceb. Biofouling in reverse osmosis membranes for seawater desalination: Phenomena and prevention. Desalination, 281 (2011) 1–16.

[8]   D. J. Miller, P. A. Araujo, P. B. Correia, M. M. Ramsey, J. C. Kruithof, M. C. Van Loosdrecht, J. S. Vrouwenvelder. Short-term adhesion and long-term biofouling testing of polydopamine and poly (ethylene glycol) surface modifications of membranes and feed spacers for biofouling control. Water Research, 46 (2012) 3737–3753.

[9]   J. S. Vrouwenvelder, S. A. Manolarakis, J. P. Van der Hoek, J. A. M. Van Paassen, W. G. J. Van der Meer, J. M. C. Van Agtmaal, M. C. M. Van Loosdrecht. Quantitative biofouling diagnosis in full scale nanofiltration and reverse osmosis installations. Water Research, 42 (2008) 4856–4868.

[10]   J. S. Vrouwenvelder, M. C. M. van Loosdrecht, J. C. Kruithof. Early warning of biofouling in spiral wound nanofiltration and reverse osmosis membranes. Desalination, 265 (2011) 206–212.

[11]   J. S. Vrouwenvelder, J. A. M. van Paassen, L. P. Wessels, A. F. Van Dam, S. M. Bakker. The Membrane Fouling Simulator: A practical tool for fouling prediction and control. Journal of Membrane Science, 281 (2006) 316–324.

[12]   J. S. Vrouwenvelder, D. G. Von der Schulenburg, J. C. Kruithof, M. L. Johns, M. C. M. Van Loosdrecht. Biofouling of spiral-wound nanofiltration and reverse osmosis membranes: A feed spacer problem. Water Research, 43 (2009) 583–594.

[13]   A. L. Ahmad, K. K. Lau, M. A. Bakar. Impact of different spacer filament geometries on concentration polarization control in narrow membrane channel. Journal of Membrane Science, 262 (2005) 138–152.

[14]   J. S. Vrouwenvelder, C. Picioreanu, J. C. Kruithof, M. C. M. Van Loosdrecht. Biofouling in spiral wound membrane systems: Three-dimensional CFD model based evaluation of experimental data. Journal of Membrane Science, 346 (2010) 71–85.

[15]   M. S. M. F. Shakaib, M. Hasani, M. Mahmood. CFD modeling for flow and mass transfer in spacer-obstructed membrane feed channels. Journal of Membrane Science, 326 (2009) 270–284.

[16]   S. S. Bucs, A. I. Radu, V. Lavric, J. S. Vrouwenvelder, C. Picioreanu. Effect of different commercial feed spacers on biofouling of reverse osmosis membrane systems: A numerical study. Desalination, 343 (2014) 26–37.

[17]   R. Valladares Linares, S. S. Bucs, Z. Li, M. AbuGhdeeb, G. Amy, J. S. Vrouwenvelder. Impact of spacer thickness on biofouling in forward osmosis. Water Research, 57 (2014) 223–233.

[18]   K. Majamaa, J. E. Johnson, U. Bertheas. Three steps to control biofouling in reverse osmosis systems. Desalination & Water Treatment, 42 (2012) 107–116.

[19] G. Massons-Gassol, G. Gilabert-Oriol, R. Garcia-Valls, V. Gomez, T. Arrowood. Development and application of an accelerated biofouling test in flat cell. Desalination and Water Treatment, 57 (2016) 23318–23325.

[20] D. Van der Kooij, H. R. Veenendaal, W. J. H. Scheffer. Biofilm formation and multiplication of Legionella in a model warm water system with pipes of copper, stainless steel and cross-linked polyethylene. Water Research, 39 (2005) 2789–2798.

[21] N. Oulahal-Lagsir, A. Martial-Gros, M. Bonneau, L. J. Blum. Ultrasonic methodology coupled to ATP bioluminescence for the non-invasive detection of fouling in food processing equipment – Validation and application to a dairy factory. Journal of Applied Microbiology, 89 (2000) 433–441.

[22] M. J. Boorsma, S. Dost, S. Klinkhamer, J. C. Schippers. Monitoring and controlling biofouling in an integrated membrane system. Desalination and Water Treatment, 31 (2011) 347–353.

Chapter 4
The importance of the feed spacer compared to the membrane in biofouling prevention

Effect of shear and permeation in biofouling development in different reverse osmosis membranes and feed spacers

Biofouling in reverse osmosis (RO) elements occurs when bacteria settle in the feed channel of an element and produce a biofilm. The biofilm evolution causes an exponential increase in the feed-concentrate pressure drop (dP) in the RO system resulting in higher energy consumption and require more frequent cleanings (CIP). Additionally, if the dP becomes too high, the elements are at risk of irreversible mechanical damage. The focus of this research was to fundamentally study biofouling and understand the influence of membrane chemistry and feed spacer. Several experimental tools were used including a Center for Disease Control (CDC) reactor, a membrane fouling simulator (MFS) and a membrane permeation flat cell (FC) unit. The results proved that the presence of the feed spacer on the membrane surface had the largest effect in promoting biofouling in RO systems. It was observed as well that the impact of the feed spacer was much more significant than the effect of the membrane chemistry (surface charge and hydrophilicity). It has been demonstrated as well that the methods and capabilities presented in this report are very reliable to evaluate new RO elements with biofouling prevention properties based on advanced membrane chemistries and feed spacer designs.

4.1 Introduction

Biofouling in RO water treatment occurs when the feed channel in the RO element is partially or fully blocked by bacteria-produced biofilm. This can cause the pressure drop (dP) across the RO element to increase, leading to a hydraulic imbalance and possibly a permanent damage of the element. Additionally, biofilms can affect membrane transport properties and create a drop in transmembrane pressure, which lowers the flux and increases the energy consumption. To regain performance and avoid element damage, the systems typically undergo a cleaning in place (CIP) using strong

Acknowledgments: The author would like to acknowledge the contributors of this chapter: Guillem Gilabert-Oriol, Gerard Massons, Tina Arrowood, Jessica Shu, Jaime Curtis-Fisk and Meaghan Woodward. The author would like to thank all DuPont Water Solutions team, and specifically Nicolas Corgnet and Javier Dewisme for their outstanding help in performing these experiments.

This chapter is in the process of being submitted for publication, and was presented in the American Dairy Science Association (ADSA) conference held in Pittsburg (United States of America) in June 2017.

https://doi.org/10.1515/9783111639079-004

basic followed by strong acidic cleaning solutions. Frequent exposure of the RO membrane polyamide to these harsh cleaning chemicals can lead to the deterioration of the membrane performance over time. Summarizing, biofouling affects the energy consumption of the system, the water productivity, the membrane lifetime, the chemicals consumptions, and ultimately the cost to produce clean water [1].

The surface biofouling mechanism described in the literature follows three main sequential phases [2, 3]. The first is the deposition of organic matter to condition the surface; the second is the attachment of functional microbial communities, and the third is the growth of a complex biofilm network composed of extracellular polysaccharides, proteins, and bacteria. In RO systems, this mechanism is affected by the hydrodynamics of an operating RO element. Three main effects influence the described mechanism. The first one is the water cross-flow that creates regions of high and low shear over the membrane surface. The second refers to water mixing from the feed spacer that creates regions of high and low shear. The third one is related to water permeation through the membrane that drives contaminants to the membrane surface.

The aim of this work was to develop testing capabilities to evaluate the design features of an RO element that impact biofilm formation. These capabilities can be used to screen new technologies and develop advanced biofouling-resistant RO elements.

4.2 Materials and methods

The laboratory biofouling test equipment employed were a Center for Disease Control (CDC) reactor [4], an MFS [5], and a membrane permeation flat cell (FC) [6]. The CDC reactor was used to compare biofilm formation on different membrane surfaces in a low shear environment. The MFS and FCs were used to evaluate biofouling under cross-flow velocities similar to an RO element and also to explore operation with and without a feed spacer. The FC units, however, operate with permeation whereas the MFS units do not. The capabilities of each of these test methods relative to a typical RO element are summarized in Table 4.1.

Table 4.1: Laboratory equipment used for studying biofouling compared to a full scale element.

Parameter	CDC	MFS	Flat cell	Element
Cross-flow velocity	No	0–0.5 m/s	0.5–1.3 m/s	0–0.3 m/s
Membrane surface	Yes	Yes	Yes	Yes
Feed Spacer	No	Yes/no	Yes/no	Yes
Permeation	No	No	Yes	Yes
Feed channel pressure drop (dP)	No	Yes	No	Yes

4.2.1 CDC reactor

The method used with the experiments undertaken in the Center for Disease Control (CDC) reactor is based on the ASTM method E2562-07. This approach involves securing membrane samples to eight rods that are placed in the CDC reactor vessel. A schematic of the CDC reactor can be observed in Figure 4.1. Membranes are placed in the reactor and exposed to a solution containing bacteria to allow for biofilm formation. A 1 mL of *Pseudomonas aeruginosa* culture grown overnight to a 10^{+8} CFU/mL is added to 350 mL of M9 media with 0.9% glucose (Sigma Aldrich). The solution is incubated at room temperature and refreshed during the experiment by flowing fresh M9 media with 0.9% glucose solution into the reactor at a rate of about 1 mL/min for about 12 h each day and collecting the overflow in a waste container. The continuous flow of fresh solution aids in maintaining a consistent level of bacteria in the reactor. The solution is stirred at 300 rpm to create a gentle shear. Samples are removed from the CDC reactor at the targeted time points. The amount of bacterial adhesion is quantified by removing samples from each membrane coupon, vortexing in cell culture media and using agar plate count to determine cell count. The shear over the membrane in the CDC reactor is significantly lower than in RO elements; however, the CDC reactor provides a means to screen up to 24 membrane chemistries at one time for high-throughput exploration.

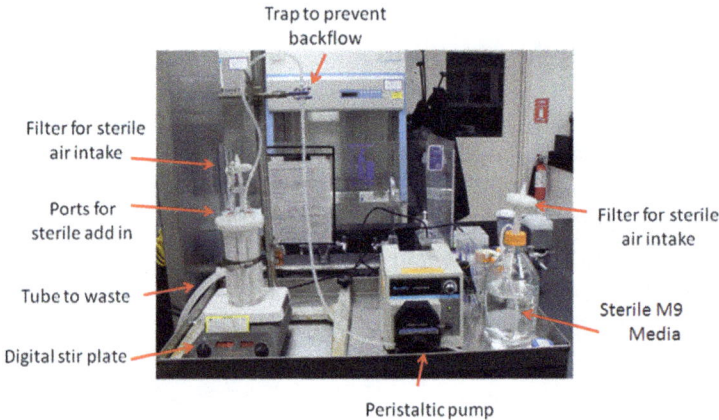

Figure 4.1: CDC reactor diagram.

4.2.2 Membrane fouling simulator

The MFS is a portable assembly that can be installed as a stand-alone test unit or in parallel with an RO system. The picture and schematic of the MFS system can be observed in Figure 4.2. These units have been described in the literature as a cost-effective way to monitor biofouling in RO systems to determine the biofouling potential of various feed waters and to compare the fouling performance of membranes and spacers [7]. The assembly includes a pressure limitation valve set at 1 bar, a flow meter, a transparent cell, a high precision pressure drop indicator, and a flow control valve. The transparent cell is assembled with a membrane and feed spacer as shown in the cross-sectional image in Figure 4.2. The MFS unit has a cross-flow velocity range that can match the feed channel velocity of a commercially operated 8-inch RO membrane element, but the MFS does not allow water permeation through the membrane. With the cross-flow velocity controlled at approximately 0.12 m/s, MFS units installed in a feed water side-stream in parallel to an industrial 8-inch RO system has shown to give similar relative differential pressure rise due to biofouling as the 8-inch systems [8]. To accelerate biofouling and enable faster screening of conditions, the feed water to the MFS can be optionally dosed with bioassimilable nutrients. This experiment was performed in the Global Water Technology Center that DuPont Water Solutions has in Tarragona, Catalonia, Spain.

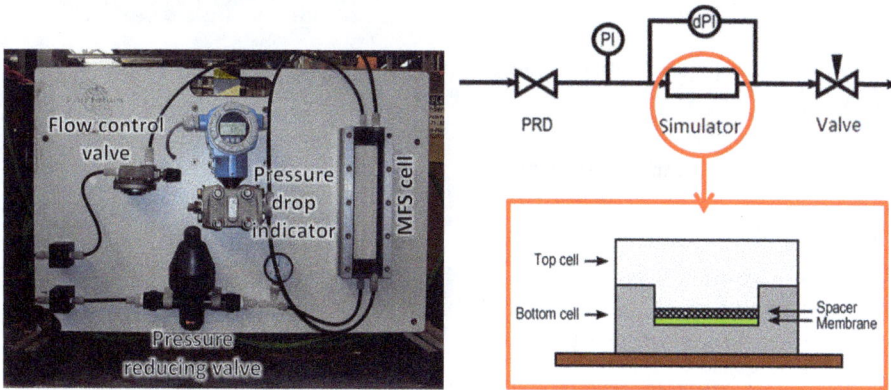

Figure 4.2: MFS simulator picture and diagram.

4.2.3 Flat cells

The membrane permeation FC units are used to closely mimic the conditions found in a spiral-wound element [9]. The system used for this study has three parallel FCs. A schematic of the FC unit, together with a picture of it can be found in Figure 4.3. These cells are assembled with a membrane and can also optionally host a feed spacer. The feed water is surface water taken from brackish water coming from Ebro River in L'Ampolla, Catalonia, Spain, after conventional pretreatment, as it is summarized in Table 4.2. The water was fed in once-through mode and can be dosed with nutrients to promote biofouling [6]. Each cell has a manual feed and concentrate flow control valve to provide equal operating fluxes (30 L/m^{2}h) for the three parallel cell assemblies. The cross-flow velocity was controlled to approximately 0.5 m/s. This is slightly higher than in an element, but is not expected to significantly impact the observations since biofouling will still develop at this higher flow rate [10]. Each test was stopped after 4 days, and the amount of biofouling was quantified by collecting a sample of the membrane and spacer and analyzing the concentration of adenosine triphosphate (ATP) and total organic content (TOC) present. This experiment was performed in the Global Water Technology Center that DuPont Water Solutions has in Tarragona, Catalonia, Spain.

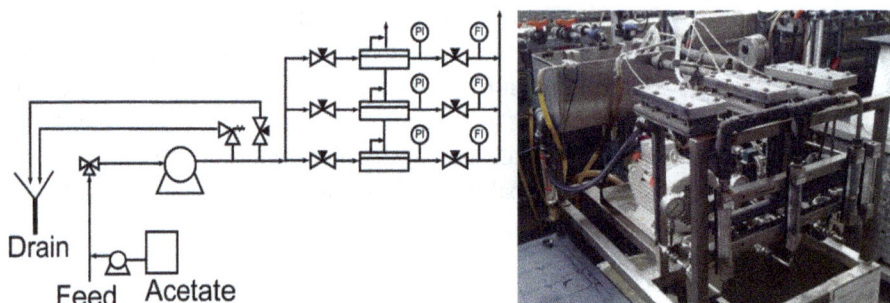

Figure 4.3: Flat cell pilot unit for flat cell testing.

4.2.4 Natural river and wastewater

Surface water pretreated with coagulation, flocculation, sand filtration from the Ebro River in L'Ampolla, Catalonia, Spain, was used in the FC experiments and was one of the two water sources used in the MFS experiments. The main composition of this water is summarized in Table 4.2.

Table 4.2: UF pretreated Ebro River water composition.

Parameter	Concentration
TDS	1,000 mg/L
TOC	1.3 mg/L
ATP	6 ng/L
Nitrate	11 mg/L
Phosphate	0.040 mg/L

The other water type used was secondary aerobic treated wastewater from Vila-Seca wastewater treatment plant in Catalonia, Spain. The water was filtered through a membrane bioreactor (MBR) and cartridge filters before feeding the MFS unit. The composition of the water to the MFS is summarized in Table 4.3.

Table 4.3: MBR pretreated Vila-Seca
wastewater characteristics.

Parameter	Concentration
TDS	1,440 mg/L
TOC	6.6 mg/L
ATP	83 ng/L
Nitrate	30 mg/L
Phosphate	0.55 mg/L

When experiments required accelerated biofouling, bioassimilable nutrients in the form of acetate, nitrate, and orthophosphate sodium salts, were dosed. The concentrations used were following the carbon, nitrogen, and phosphorous ratios described in the literature [11, 12]. Three different nutrient concentrations were used in order to study the effect on biofilm development. The highest concentration provided 0.40 mg/L carbon, 0.08 mg/L nitrogen, and 0.04 mg/L phosphate; the intermediate concentration provided 0.20 mg/L carbon, 0.04 mg/L nitrogen, and 0.02 mg/L phosphate; and the lowest provided 0.10 mg/L carbon, 0.02 mg/L nitrogen, and 0.01 mg/L phosphate additional nutrients to the feed water.

4.2.5 ATP and TOC analytical methods to quantify biofouling

In addition to monitoring the pressure drop increase of the MFS cells during the experiment, a sample of the membrane and spacer was analyzed after the experiment to quantify the level of ATP and TOC present. These analyses provide a measure of the bacteria and biofilm present in the feed channel. ATP and TOC were used as the exclusive measures for biofouling in the FCs where no dP measurements were available.

The membrane and feed spacer sample materials were analyzed by cutting a 4 × 4 cm square sample of membrane and spacer from the center of the exposed area. The sample was then placed into 20 mL of ultrapure water and sonicated at room temperature for three intervals of 2 min each.

The ATP content of the water was analyzed using a Celsis Advance Luminometer. The response is based on luciferin bioluminescent reaction with ATP, which provides a light signal. This light signal is analyzed with a photo-detector, and it is later converted to ATP concentration. This parameter is proportional to the amount of living microorganisms present in the sample and can be directly correlated to colony-forming units (CFUs) [13].

Additionally, TOC analyses were also performed using the same fouling extract method described above. TOC content in liquid samples was analyzed using a TOC-L Shimadzu based on the UNE EN-1484:1998 method. The sample was oxidized via high temperature catalytic combustion and quantified using an infrared detector.

The conversion was performed using a calibration curve obtained from a TOC standard solution. The organic carbon concentration is proportional to the amount of organic material present in the sample like extracellular polymeric substances (EPSs) [14].

4.2.6 Membranes and feed spacers tested

Smoother, more hydrophilic, and lower surface-charged membranes are described in the literature as being more fouling resistant [2, 15–17]. Therefore, two membranes, membrane A and membrane B, were chosen in this study to represent examples of a low and high fouling resistance membrane surface, respectively. The properties of the two membranes are summarized in Table 4.4. Membranes A and B have similar degrees of roughness, but membrane A is more hydrophobic and more negatively charged than membrane B and is expected to be fouled more rapidly. Standard polypropylene extruded feed spacers were used in the MFS and FC experiments. The material had 9 strands/inch and was either 28-mil feed spacer or 34-mil feed spacer thick (1 mil = 0.001 inch). The values for the atomic force microscopy (AFM), the contact angle, and the zeta potential are displayed in Table 4.4.

Table 4.4: Membrane characteristics.

Parameter	Membrane A	Membrane B
AFM average roughness	61 nm	61 nm
Contact angle	72°	43°
Zeta potential (pH 8)	−25 mV	−5 mV

4.3 Results and discussion

The biofouling testing equipment described above provides the flexibility to evaluate the effect of several parameters on membrane chemistry to resist biofilm formation. The variables studied are the impact of shear, membrane flux, presence of a feed spacer, and concentration of bioassimilable nutrients. This fundamental understanding not only provides direction for designing fouling resistant elements in the future, but also provides effective laboratory screening tools to evaluate new designs prototypes. Table 4.5 summarizes the experimental design employed in this study and the results of each are discussed herein.

Table 4.5: Summary of the experiments performed.

Test	Asset	Screening throughput	Cross-flow	Flux	Spacer	Nutrients
3.1	CDC	High	No	No	No	High
3.2	CDC	High	No	No	No	Low
3.3	MFS	Mid	Typical	No	Yes	River water
3.4	MFS	Mid	Typical	No	Yes/no	Wastewater with nutrients
3.5	FC	Low	High	Yes	Yes/no	River water with nutrients

4.3.1 No shear and high-nutrient dosage without spacer or permeation (CDC)

The CDC Reactor is an especially attractive technique for exploring biofilms since it provides the potential to screen multiple surface chemistries within a short period of time. In this work, however, the aim was to first demonstrate that this method could easily differentiate two different membrane surface types, A and B, on the basis of biofouling accumulation.

Figure 4.4 shows the evolution of the CFUs on the surface of membrane A and membrane B during a 2-h study using a high bacteria concentration in the CDC reactor. In this study, the fouling solution was prepared by adding 1 mL of a high cell density cell culture to the 350 mL CDC reactor fouling unit. The goal was to achieve a bacteria concentration of approximately 10^6 CFU/mL. Membrane samples were then taken and analyzed after 30, 60, 90, and 120 min with the expectation that the CFUs on the membrane would gradually increase over this time period. The results, however, showed that bacteria initially attached very quickly and did not increase over the course of the 120 min, as can be observed in Figure 4.4. Additionally, both membranes, despite the different surface properties, gave similar results.

The results raise the question if the high bacteria concentration in the CDC reactor provides fouling conditions that were too accelerated. After just 30 min, there was substantial adhesion of bacteria on both membrane surfaces, while reaching these levels of bacteria in typical RO applications occurs over the course of weeks to months [18].

4.3.2 No shear and low-nutrient dosage without spacer and permeation (CDC)

To evaluate the effect of slower evolution of biofouling in the CDC reactor the experiment was repeated using a lower bacteria concentration. The number of bacteria was reduced to approximately 10^4 CFU/mL. The fouling process was also slowed by limiting the concentration of bacteria in the CDC reactor by flowing sterile nutrient media into the reactor. The goal was to displace the established cell culture with the addition of fresh solution at a rate of about 1 mL/min for about 12 h each day. Membrane samples were removed and analyzed after 2, 7, 9, and 14 days. Results are shown in Figure 4.5.

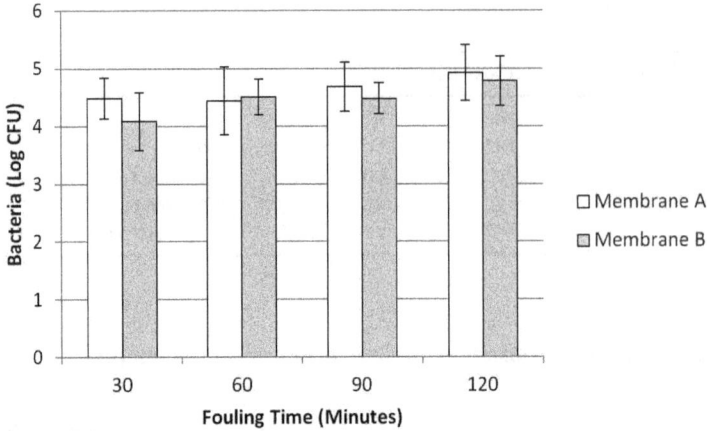

Figure 4.4: CFU evolution of different membranes in an accelerated trial on CDC.

This data shows a more gradual development of biofilm on the surface of the membrane over the course of the experiment, but again there were no significant differences observed between membranes A and B.

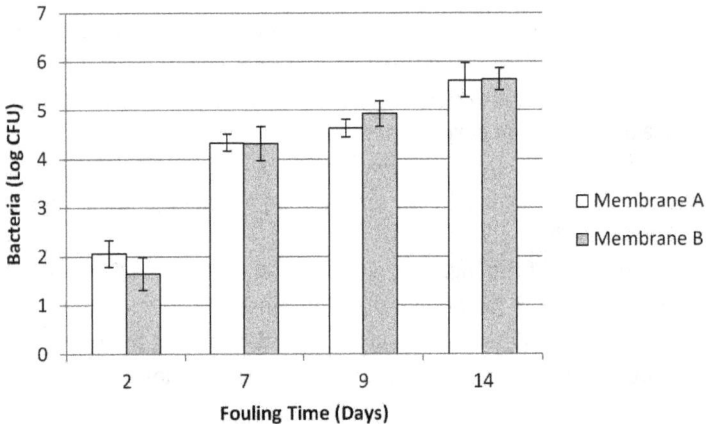

Figure 4.5: CFU evolution of different membranes in a slower trial on CDC.

The CDC reactor provides a high-throughput means to evaluate various membrane surfaces, but under the test conditions no difference in the two membranes was observed. Based on these results, membrane chemistry might not be enough to provide a strong biofouling resistance. The CDC reactor does not capture the dynamic biofouling environment of an RO element; thus additional methods were explored.

4.3.3 Typical shear and spacer without permeation (MFS)

Bacteria are known to express differently under high and low shear environments [10]. The water cross-flow within an RO element feed channel provides much higher shear environment than the CDC reactor. Additionally, the feed spacer in the feed channel provides water mixing, which creates regions of high and low shear. The MFS test unit provides a means to evaluate membrane under the typical water cross-flow environments of an RO element. Four portable MFS units were assembled in parallel and connected to the brackish water source characterized in Table 4.2. Each MFS unit had a membrane and a 34-mil feed spacer. Two units were assembled with membrane A and two with membrane B to provide replicates for each membrane type.

4.3.3.1 ATP and TOC results
After 5 days of operation the pressure drop of the MFSs increased from 0.05 bar to 0.14 bar. The units were removed from service, disassembled, and membrane and spacer samples were analyzed for ATP and TOC. The results summarized in Table 4.6 show that the two membrane types could not be distinguished with statistical relevance. However, the average data suggests that membrane B had slightly less biofilm accumulation than membrane A.

Table 4.6: Analytical results for typical shear with 34-mil spacer on MFS.

ATP (ng/cm^2)		TOC (mg/m^2)	
Membrane A	**Membrane B**	**Membrane A**	**Membrane B**
4.1 ± 2.1	2.7 ± 0.6	47.2 ± 11.6	40.3 ± 3.3

4.3.3.2 Visual observations
The transparent window of the MFS cells allowed addition visual observations. Figure 4.6 includes a series of photographs of the cells containing membrane A and membrane B, where the flow direction in each photo is from left to right, and where the inlet position and center positions of the coupons are magnified. Two important observations could be extracted from these pictures. Firstly, more biofilm appeared close to the feed inlet side of the cell than the center. This is in line with what is reported in the literature – that biofilm mainly develops in the lead elements when RO elements are installed inside a pressure vessel [19–22]. Secondly, the biofilm in the form of bulk slime appeared to be more localized on the feed spacer than on the membrane. In particular, the biofilm is located at the spacer strand intersections where zones of low shear are present. It should be noted that the observations shown in Figure 4.6 were also observed in the replicate set of cells.

Membrane A (Front)   Membrane A (Center)

Membrane B (Front)   Membrane B (Center)

Figure 4.6: Biofilm distribution on the membrane surface and the 34-mil spacer on MFS.

Biofilm is highly hydrated and can be difficult to be observed on the white membrane background. Therefore, in order to visually examine the biofilm on top of the membrane surface, each MFS cell was opened at the end of the experiment. The feed spacer was removed, and biofouling specific stain was poured onto the membrane surface to provide a contrast between the biofilm and the membrane surface. Bacteria create three dimensional structures when they build biofilms. The stain helps provide a visual contrast between the clean membrane surface (white) and the biofilm inner structure (black). Figure 4.7 shows the visual comparison of both membrane types of surfaces using this technique. Membrane B has fewer black spots and thus less biofilm. This difference was not observed in the ATP and TOC analytical data due to these measurements including the total biofilm that occurs on both the membrane's surface and feed spacer. The regions where feed spacer was present is in particular are where biofilm appear to be most dense. These are regions of water flow around the feed spacer where there is less turbulence and could serve as regions for bacteria to settle and colonize to form the biofilm. When comparing feed spacer photos shown in Figure 4.6, together with the membrane surface photos, as shown in Figure 4.7, it could be seen that visually less biofilm developed on the membrane surface than in the feed spacer. This highlights the important role that feed spacers have in biofouling formation.

Membrane A (Center)   **Membrane B (Center)**

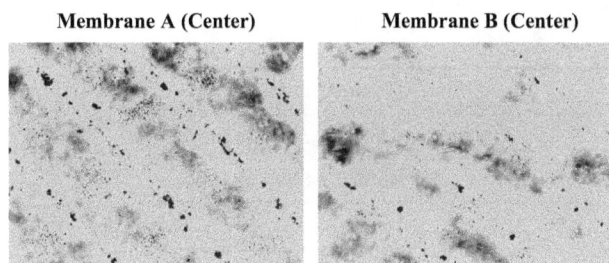

Figure 4.7: Biofilm distribution on membrane surface after removing the 34-mil spacer on MFS.

4.3.4 Typical shear with and without spacer and without permeation (MFS)

The results in the previous experiment show the dominant effect that the feed spacer has on accumulating biofilm, and it masks the effect of the different membrane types. Conducting the cross-flow experiment in the absence of the feed spacer was done in an attempt to observe only the membrane effect. Four MFS units were used in this experiment – two contained membrane A and two contained membrane B. One of each pair was equipped with a 34-mil feed spacer and the other had a modified 34-mil feed spacer where the feed spacer in the center of the cell was removed but the inlet and outlets had small pieces of feed spacer used as shims to maintain the feed channel height of the MFS. The experiment was concluded after 7 days at which time the pressure drop of the vessels hosting the standard 34-mil feed spacer increased from 0.10 bar to 0.43 bar.

4.3.4.1 ATP and TOC results

Samples of the membrane and feed spacer (if present) were cut from the center portion of the coupon and analyzed for ATP and TOC. Results are summarized in Table 4.7. The two cells fitted with the 34-mil feed spacer gave similar results to the previous experiment. Membranes A and B were difficult to distinguish, especially since there were no replicates in this experiment. However, the two cells that did not have a feed spacer in the center of the MFS showed slightly lower values for membrane B than for membrane A. The most relevant observation is that, despite all four MFS units operating with the same feed flow rates, those with the modified feed spacer have significantly less biofilm in the center of the MFS unit than those with the standard 34-mil spacer. This might be due to the presence of low shear zones that the feed spacer creates across the feed-concentrate channel, which may facilitate biofilm attachment. In the absence of a feed spacer, the low shear (dead) zones are eliminated.

Table 4.7: Analytical results with and without 34-mil spacer on MFS.

ATP (ng/cm^2)				TOC (mg/m^2)			
Membrane A		Membrane B		Membrane A		Membrane B	
34 mil	No spacer	34 mil	No spacer	34 mil	No spacer	34 mil	No spacer
19.3	2.2	21.4	1.2	56.3	14.3	44.8	10.5

4.3.4.2 Visual observations

In addition to the analytical results shown in Table 4.8, photos of the MFS cells with and without 34-mil feed spacer were taken. This can be observed in Figure 4.8. Both photos show that with the feed spacer present, more biofilm can be observed in the MFS cell. As with the previous experiment, more biofilm is seen at the feed side (left) of the MFS cell than at the center. This is also observed in the experiments with the modified feed spacer where brown biofilm is present in the short section of feed spacer at the inlet, but the center of the cell appears very clean.

MFS cell with 34 mil spacer   MFS cell without spacer

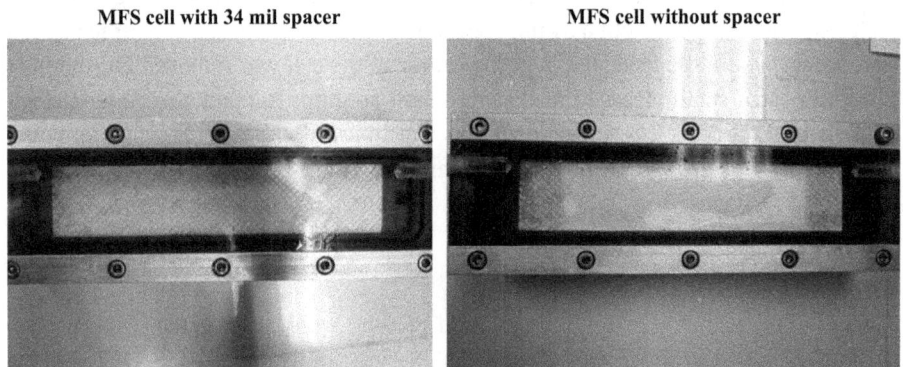

Figure 4.8: Biofilm distribution with and without 34-mil spacer in the MFS cells.

The MFS units were disassembled and the feed spacer removed to visualize the biofilm on the membrane using the stain. The center exposed area of the membranes was compared. Membrane A and membrane B paired with the standard 34-mil feed spacer provided no observable difference. There was also no difference observed between membranes A and B when each was paired without feed spacer. However, comparing membrane A with the standard feed spacer with the one without spacer, significantly more biofilm is present on the membrane surface when the standard feed spacer was present, as it can be observed in Figure 4.9. This is in agreement with the visual observation shown in Figure 4.8 and the analytical results in Table 4.7.

Membrane where spacer is used Membrane where no spacer is used

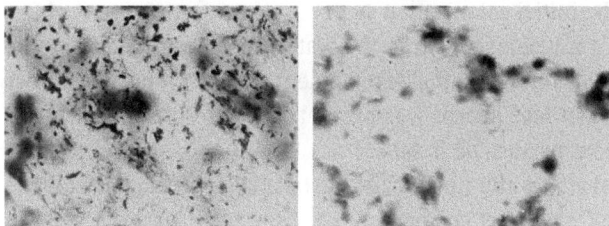

Figure 4.9: Membrane surfaces comparison where 34-mil spacer and no spacer are used on MFS.

4.3.5 High shear with and without spacer and permeation (flat cell)

The CDC and MFS results show little differentiation of membrane A and membrane B in terms of biofilm development. However, the MFS results indicated a slight advantage for membrane B. One explanation for the lack of differentiation may be the absence of permeation in the cross-flow experiments. Permeation draws water contaminants to the membrane surface; thus membranes that are more organic fouling resistant will have a slower rate of forming the organic conditioning layer, which is the first step in biofouling [23]. FC units provide the ability to evaluate the impact of water permeation in addition to feed channel hydrodynamics on the fouling performance of different membrane surfaces.

The FC unit available for this study is limited to three cells operating in parallel. These three cells are fed with a common feed that is either dosed with 0.1, 0.2, or 0.4 mg/L carbon nutrients. Membrane A was installed in two of the cells and membrane B in the third. The cell with membrane B and one with membrane A was also equipped with a feed spacer. The second membrane A cell was operated without a feed spacer. The first four sets of experiments evaluating different nutrient dosing concentrations were conducted with 28-mil feed spacer and the last two sets of experiments with 34-mil feed spacers. Each test was stopped after 4 days and samples of the membrane and the spacer, if present, of each FC unit were analyzed for ATP and TOC concentrations, as is summarized in Tables 4.8 and 4.9, respectively.

4.3.5.1 ATP and TOC results
Comparing the ATP and TOC results of membrane A and membrane B operated in the presence of a feed spacer shows membrane B accumulates less biofouling regardless of the nutrient dosing level, as summarized in Table 4.8. In this experiment the 28-mil feed spacer was used, the exception being experiment 3 where the ATP value was

slightly higher for membrane B (this may be an outlier). The differentiation between the membrane types was more pronounced in experiments 1 and 2 when more biofilm was formed due to the higher nutrient addition. These results suggest that the membrane surface properties have more influence on the evolution of the biofilm on its way to maturation than the initial colonization step.

Regardless of the level of nutrients added to the water, the results show the important role the feed spacer has on biofouling development. The ATP and TOC values were always much higher when feed spacer was present. These results are in agreement with previous experiment performed in MFS.

As expected, the ATP levels on the membrane surface are correlated with the level of nutrients added. This is a direct correlation when the feed spacer is present. However, in the absence of the feed spacer there appears to be a critical threshold between 0.2 mg/L and 0.4 mg/L C nutrient dosing levels, which trigger the bacteria to colonize and multiply).

Table 4.8: Analytical results with and without 28-mil spacer on flat cells.

Test	C (mg/L)	ATP (ng/cm^2)			TOC (mg/m^2)		
		Membrane A		Membrane B	Membrane A		Membrane B
		28 mil	No spacer	28 mil	28 mil	No spacer	28 mil
1	0.4	239.4	48.4	170.1	199.4	20.2	166.8
2	0.2	109.3	4.0	42.7	64.6	22.8	51.5
3	0.1	23.9	2.4	26.6	37.5	11.9	31.4
4	0.1	8.6	0.4	6.9	53.9	9.3	41.0

4.3.5.2 Visual observations

Figure 4.10 shows a comparison of the visual observation of the biofilm on membrane A and membrane B exposed with a 28-mil feed spacer (right and left pictures) to membrane A exposed without a feed spacer present (center). The brown coloration of the biofilm slime produced is evident only on the membranes exposed with a feed spacer present. This photo was obtained in experiment 1, but is representative of all the other experiments performed.

The photos in Figure 4.11 show the visual comparison of the stained membrane after the feed spacer is removed from Experiment 4. Membrane A showed more biofouling than membrane B when tested with a 28-mil feed spacer, but in the absence of the feed spacer, membrane A had much less biofilm.

Membrane A
(28 mil)   Membrane A
(no spacer)   Membrane B
(28 mil)

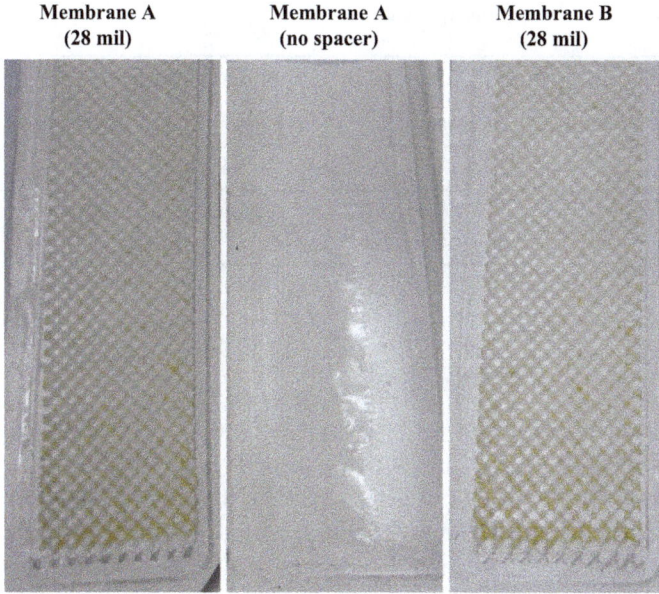

Figure 4.10: Biofouling developed on the membrane surface and feed spacer on flat cells.

Membrane A
(28 mil)   Membrane B
(28 mil)   Membrane A
(No spacer)

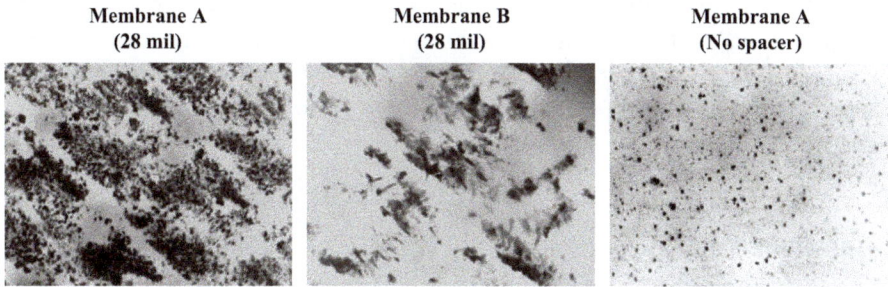

Figure 4.11: Biofouling developed on the membrane surface on flat cells.

4.3.5.3 Experiment with 34-mil spacer

The results from the two FC experiments conducted with 34-mil feed spacers instead of the 28-mil spacers supported the same conclusions. Very little fouling occurred in the absence of the feed spacer, but significantly high amount of fouling is obtained when the feed spacer was present. This is consistent with the hypothesis exposed in the literature that the best spacer is the membrane that has no spacer, in order to prevent biofouling in RO systems [24].

Table 4.9: Analytical results with and without 34-mil spacer on flat cells.

Test	C (mg/L)	ATP (ng/cm^2)			TOC (mg/m^2)		
		Membrane A		Membrane B	Membrane A		Membrane B
		34 mil	No spacer	34 mil	34 mil	No spacer	34 mil
5	0.2	66.9	16.7	38.4	88.3	26.3	49.0
6	0.2	97.5	30.4	42.3	33.6	10.5	15.8

4.4 Conclusions

In this study, two membranes expected to have low (membrane A) and high (membrane B) fouling resistance were examined under various biofouling conditions. Biofouling developed in both membranes. However, membrane B showed slightly improved biofouling resistance but only when exposed to moderate to high biologically active water, with cross-flow over the membrane surface and in the presence of a feed spacer. The feed spacer is shown to have the biggest impact on promoting biofouling compared to the membrane surface, as when the spacer was not present, biofouling hardly developed in cross-flow systems. Under these conditions, both membranes A and B exhibited very little biofouling. The MFS and flat cell methods described in this study are found to be the most reliable and will be used to screen improved feed spacer designs and fouling resistant membrane chemistries for advancing the product line of DuPont™ FilmTec™ RO membranes, which are fouling-resistant.

References

[1]   A. Matina, Z. Khana, S. M. J. Zaidia, M. C. Boyceb. Biofouling in reverse osmosis membranes for seawater desalination: Phenomena and prevention. Desalination, 281 (2011) 1–16.
[2]   M. Herzberg, M. Elimelech. Biofouling of reverse osmosis membranes: Role of biofilm enhanced osmotic pressure. Journal of Membrane Sciences, 295 (2007) 11–20.
[3]   E. Bar-Zev, I. Berman-Frank, O. Girshevitz, T. Berman. Revised paradigm of aquatic biofilm formation facilitated by microgel transparent exopolymer particles. Proceedings of the Nation Academy of Sciences, 109 (2012) 9119–9124
[4]   ASTM E2562-12. Standard Test Method for Quantification of Pseudomonas aeruginosa Biofilm Grown with High Shear and Continuous Flow using CDC Biofilm Reactor, ASTM International, West Conshohocken, PA. (2012). www.astm.org.
[5]   J. S. Vrouwenvelder, J. A. M. van Paassen, L. P. Wessels, A. F. van Dam, S. M. Bakker. The membrane fouling simulator: a practical tool for fouling prediction and control. Journal of Membrane Science, 281(1–2) (2006316–324

[6]   G. Massons-Gassol, G. Gilabert-Oriol, R. Garcia-Valls, V. Gomez, T. Arrowood. Development and application of an accelerated biofouling test in flat cell. Desalination and Water Treatment, 57(48–49) (201623318–23325

[7]   J. S. Vrouwenvelder, S. M. Bakker, M. Cauchard, R. Le grand, M. Apacandie, M. Idrissi, S. Lagrave, L. P. Wessels, J. A. M. Van paassen, J. C. Kruithof, M. C. M. van Loosdrecht. The membrane fouling simulator: a suitable tool for prediction and characterisation of membrane fouling. Water Science and Technology, 55(8–9) (2007197–205

[8]   G. Massons-Gassol, G. Gilabert-Oriol, J. Johnson, T. Arrowood. Comparing biofouling development in membrane fouling simulators and spiral-wound reverse osmosis elements using river water and municipal wastewater. Industrial & Engineering Chemistry Research, 56(40) (201711628–11633

[9]   L. Vanysacker, P. Declerck, I. Vankelecom. Development of a high throughput cross-flow filtration system for detailed investigation of fouling processes. Journal of Membrane Science, 442 (2013) 168–176.

[10]   J. S. Vrouwenvelder, C. Hinrichs, W. G. J. Van der meer, M. C. M. van Loosdrecht, J. C. Kruithof. Pressure drop increase by biofilm accumulation in spiral wound RO and NF membrane systems: role of substrate concentration, flow velocity, substrate load and flow direction. Biofouling, 25 (2009) 543–555.

[11]   J. D. Jacobson, M. D. Kennedy, G. Amy, J. C. Schippers. Phosphate limitation in reverse osmosis: An option to control biofouling? Desalination and Water Treatment, 5 (2009) 198–206.

[12]   J. S. Vrouwenvelder, M. C. M. Van Loosdrecht. Phosphate limitation to control biofouling. Water Research, 44 (2010) 3454–3466.

[13]   S. E. Jensen, P. Hubrechts, B. M. Klein, K. R. Hasløv. Development and validation of an ATP method for rapid estimation of viable units in lyophilised BCG Danish 1331 vaccine. Biologicals, 36 (2008) 308–314.

[14]   R. J. Barnes, R. R. Bandi, F. Chua, J. H. Low, T. Aung, N. Barraud, A. G. Fane, S. Kjelleberg, S. A. Rice. The roles of Pseudomonas aeruginosa extracellular polysaccharides in biofouling of reverse osmosis membranes and nitric oxide induced dispersal. Journal of Membrane Science, 466 (Sep 2014), 161–172.

[15]   M. Herzberg, S. Kang, M. Elimelech. Role of extracellular polymeric substances (EPS) in biofouling of reverse osmosis membranes. Environmental Science & Technology, 43 (2009) 4393–4398.

[16]   E. M. Van wagner, A. C. Sagle, M. M. Sharma, Y. H. La, B. D. Freeman. Surface modification of commercial polyamide desalination membranes using poly (ethylene glycol) diglycidyl ether to enhance membrane fouling resistance. Journal of Membrane Science, 367 (2011) 273–287.

[17]   E. Matthiasson. The role of macromolecular adsorption in fouling of ultrafiltration membrane. Journal of Membrane Science, 16 (1983) 23–36.

[18]   L. A. Bereschenko, H. Prummel, G. J. W. Euverink, A. J. M. Stams, M. C. M. van Loosdrecht. Effect of conventional chemical treatment on the microbial population in a biofouling layer of reverse osmosis systems. Water Research, 45 (2011) 405–416.

[19]   J. S. Vrouwenvelder, M. C. M. Van Loosdrecht, J. C. Kruithof. Early warning of biofouling in spiral wound nanofiltration and reverse osmosis membranes. Desalination, 265(1) (2011206–212

[20]   M. B. Dixon, S. Lasslett, C. Pelekani. Destructive and non-destructive methods for biofouling analysis investigated at the Adelaide Desalination Pilot Plant. Desalination, 296 (2012) 61–68.

[21]   J. S. Vrouwenvelder, J. C. Kruithof, M. C. M. Van Loosdrecht. Integrated approach for biofouling control. Water Science & Technology, 62 (2010) 2477.

[22]   J. S. Vrouwenvelder, S. A. Manolarakis, J. P. Van der hoek, J. A. M. van Paassen, W. G. J. van der Meer, J. M. C. Van agtmaal, H. D. M. Prummel, J. C. Kruithof, M. C. M. van Loosdrecht. Quantitative biofouling diagnosis in full scale nanofiltration and reverse osmosis installations. Water Research, 42 (2008) 4856–4868.

[23] S. R. Suwarno, X. Chen, T. H. Chong, V. L. Puspitasari, D. McDougald, Y. Cohen, S. A. Rice, A. G. Fane. The impact of flux and spacers on biofilm development on reverse osmosis membranes. Journal of Membrane Science, 405 (2012) 219–232.

[24] E. M. Hoek, T. M. Weigand, A. Edalat. Reverse osmosis membrane biofouling: causes, consequences and countermeasures. Npj clean water, 5(1) (202245

Chapter 5
Effect of temperature and development of quick biofouling test

Development and application of an accelerated biofouling test in flat cell

This chapter discusses the steps followed to establish and statistically validate an accelerated biofouling creation protocol by dosing nutrients in the flat cell experimental unit. This method is statistically validated using adenosine triphosphate (ATP) and total organic carbon (TOC) analysis from membrane and feed spacer. This happens as pressure drop measurement is not available and flux loss and salt rejection do not change over time. The amount of biofouling present is found to be temperature-dependent. Therefore, for a temperature above 17 °C, a 3-day test with a nutrient's dosage of 0.2 mg/L carbon dosage is used. For colder water temperatures, a 5-day test with a nutrients dosage of 1.0 mg/L carbon is developed. The reproducibility of the method is in the same range as other biofouling protocols found in the bibliography. The ultimate goal is to use this method to obtain a high throughput method to quickly screen biofouling resistance of different membrane types and feed spacers. Additionally, it can be observed how biofouling development is strongly dependent on the availability of bioassimilable nutrients.

5.1 Introduction

Reverse osmosis membranes are prone to fouling due to the trace contaminants found in natural feed water [1]. The term "fouling" in RO includes the accumulation of material on the membrane surface and/or feed spacer. If this phenomenon is not addressed, the element could suffer from a severe loss of performance [2] There are four main types of fouling in the reverse osmosis membranes, including colloidal, biological, organic, and scale (precipitated inorganic salts). Of these, biological fouling is characterized to be challenging for prevention and control [3].

Acknowledgments: The author would like to acknowledge the contributors of this chapter: Gerard Massons, Guillem Gilabert-Oriol, Tina Arrowood. The authors would like to acknowledge all the DuPont Water Solutions team for their support in this research. Especially thanks for Miguel Mestres for the technical advising and Veronica Gomez and Carolina Martinez de Peon for all the analytical support.

This chapter was originally published with the following reference: G. Massons-Gassol, G. Gilabert-Oriol, R. Garcia-Valls, V. Gomez, T. Arrowood, Development and application of an accelerated biofouling test in flat cell. Desalination and Water Treatment, 57 (2016) 23,318–23325.

https://doi.org/10.1515/9783111639079-005

Laboratory experimental methods are needed to more rapidly and systematically design RO element parameters that may impact biofouling, such as membrane chemistry and feed spacer configuration. Current published methods for reproducing biofilm on thin-film composite polyamide-based membranes are based on bacteria attachment determination [5]. The main protocols applied are the immersion test using the Center for Disease Control (CDC) biofilm reactor [6] and filtration with high concentration of bacterial solution [7, 8, 19]. Although these methods are commonly used, they are not realistic when simulating reverse osmosis conditions. The potential dependence of feed pressure, cross-flow velocity, feed spacer hydrodynamics, and/or feed water composition on the results is not well understood and, at times, these variables are not measured or controlled. Therefore, it remains challenging to correlate the data obtained using these methods with operational data [3]. Moreover, these methods require more sophisticated laboratory equipment and safety standards as they involve bacteria culturing.

The flat cell unit is a commercial membrane testing apparatus commonly used to measure flux and rejection of water treatment membranes. It does so by "sandwiching" a sample of flat sheet membrane between two plates. One plate hosts a water inlet (feed) and an outlet (concentrate) port connected by a water-feed channel. The other plate has an outlet channel for water permeating the membrane.

Under known feed flow, feed pressure, and feed composition, the flux and rejection performance of different membrane flat sheet can be measured using the flat cell. The goal of the work in this study is to develop a protocol using flat cell unit to enable one to explore the impact of different membrane chemistries on preventing biofouling. This method can provide a performance screen without needing the extra investment of building complete RO modules.

5.2 Materials and methods

5.2.1 Flat cell description

This experiment has been carried out in the Global Water Technology Center that DuPont has in Tarragona, Catalonia, Spain. The flat cell experimental plant has three side-by-side flat cells. The unit can be manually operated in recirculation mode using a cooling jacket for temperature control or in once-through mode, where there is no temperature control capability, as shown in Figure 5.1. Feed spacer and membrane leaf are manually cut using a specific template and a box cutter to fit the cell. The cell details are depicted in Figure 5.2. Each cell has an active area of 84 cm^2 (13 inch^2) and the feed spacer is placed in the bottom cell void space. The O-rings ensure the system is watertight, as observed in Figure 5.2. The measured void space in the cell is 31 mil. This channel height is constant along the cell length. A concentrated nutrient solution (labeled "Acetate" in Figure 5.1) is dosed from a tank to the feed water in order to accelerate the fouling process. Feed pressure and feed flow are adjusted using a by-pass needle valve. Individual feed flows and recoveries are set through adjusting the feed and concentrate valves. Feed flows are measured using individual flowmeters, with a range from 1 to 4 L/min (15–63 gph).

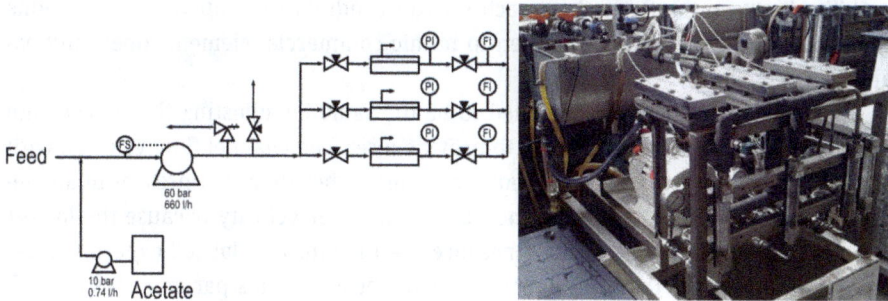

Figure 5.1: Flat cell unit for flat cell testing.

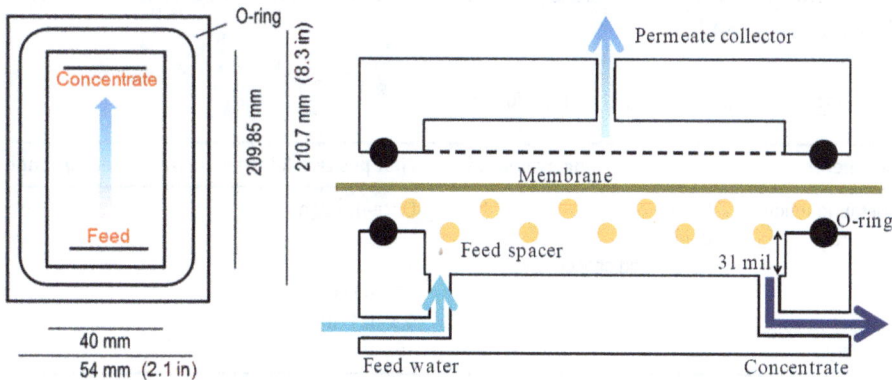

Figure 5.2: Flat cell sketch and cross-sectional configuration.

Some upgrades to the system are implemented to achieve the required operational control. These include replacing pressure and flow indicators for higher precision measurements, establishing a sanitization protocol to clean the unit after each bio-fouling experiment, installing a nutrient dosing pump and calibrating it using back-pressure, and measuring cell void spacer to calculate the cross-flow velocity. Sanitiza-tion is done by a two-step protocol. Firstly, a caustic recirculation is performed. Secondly, the system is soaked overnight using a new cleaning solution.

5.2.2 Operating conditions

The system is operated with the three flat cells operated in parallel in once-through mode. Each experimental run documented in this study uses the same condition in each of the three flat cells to provide triplicate data points and a measure for repro-ducibility between the flat cells at each set run condition. The operating conditions for the biofouling protocol are chosen to mimic commercial elements operating pa-rameters most closely.

The flux for each cell is controlled to be the same by adjusting the concentrate pressure. The feed flow is controlled to provide a set feed channel flow velocity, with a lower limit of 0.6 m/s, which is about five times higher than that present in an ele-ment. It is not possible to operate the flat cell at lower velocity because the lowest flows that the plant flowmeters can measure are 1 L/min. The flat cell experiment re-covery is much lower than the typical element recovery. This parameter is also af-fected by the measuring range of the flowmeters if the membrane is operated at a realistic flux. The comparison of the operating range evaluated, with the typical oper-ating ranges of a first position 8-inch spiral wound-reverse osmosis membrane, is shown in Table 5.1 [20].

Table 5.1: Quick biofouling method protocol for flat cell.

Parameter	Operating range evaluated	First position 8-inch element operating range
Operating mode	Once-through	Once-through
Membrane	BW30	
Spacer	28-mil feed spacer	
Flux	58–20 LMH	27–20 LMH
Flow velocity	1.2–0.6 m/s	0.3 –0.1 m/s
Recovery	0.3–0.2%	15–10%

Cross-flow velocities (v) are calculated using eq. (5.1), where v is the velocity, Q is the average feed flow and S is the cross-sectional area also known as void space. Feed-concentrate channel cross section is defined by the feed spacer thickness in the ele-ment and the feed channel height of the flat cell.

Cross-flow velocity

$$v = \frac{Q}{S}$$

(5.1)

5.2.3 Flat sheet membranes

DuPont FilmTec™ BW30 membrane sheet is selected as the reference membrane chemistry for the development of biofouling methods. This membrane is a well-known membrane that has been vastly used in a wide variety of cases. With a minimum salt rejection on 99.0%, a stabilized salt rejection of 99.5% and a pH resistance from 1 to 13, this membrane offers reliable performance and robustness across a wide range of feed conditions.

5.2.4 Feed spacers

The feed spacers used are defined by their stands per inch, thickness, and angle, and are made from polypropylene. Feed spacers are oriented in the flat cell, so the feed flow intersects the angle formed by the strands knot. The feed spacers create a void space within the membrane leaves, allowing the water to pass from the feed to the concentrate side of the element. The thicker feed spacer, results in a more open feed channel. If different feed spacer thicknesses are operated at the same feed flow, the thinner feed spacer configuration will have a higher cross-flow velocity. Cross-flow velocity affects the shear stress created by the water flow and biofilm formation [9]. Details of the two spacers used in this study, the 22-mil feed spacer, and the 28-mil feed spacer, are summarized in Table 5.2. The visual images of both the 22-mil spacer and the 28-mil spacer can be found in Figure 5.3.

Table 5.2: Details of the spacers used.

Spacer	Thickness (inch)	Strands/in	Angle (°)
22 mil	0.022	9	90
28 mil	0.028	9	90

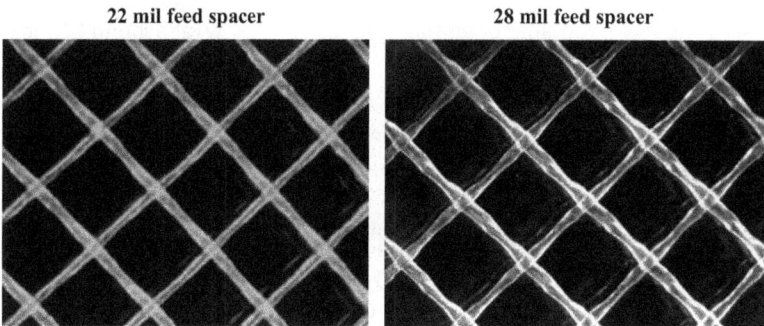

22 mil feed spacer   28 mil feed spacer

Figure 5.3: Overview of the two spacers used with the 22 mil on the left and the 28 mil on the right.

5.2.5 Brackish water characterization

The water used to feed the flat cell unit while operated in once-through mode is brackish water taken from Ebro River in L'Ampolla, Catalonia, Spain. The main pre-treatment steps are coagulation, flocculation, sand filtration, and chlorination, as shown in Table 5.3. After its treatment, water is transported 100 km to Tarragona, where it is distributed to the industrial chemical complex through Aguas Industriales de Tarragona S.A. (AITASA) company.

Table 5.3: River water pretreatment.

Step	Treatment
1	Chlorine dioxide and coagulant addition
2	Polyelectrolyte addition
3	Flocculation
4	Lamellar decantation
5	Sand bed filtration
6	Granular activated carbon filtration
7	Post chlorination

Water composition most often linked to biofouling includes phosphate (PO_4), nitrate (NO_3), TOC, and ATP. The concentration of each of these contaminants, in addition to total dissolved solids (TDS), is monitored over the course of a year in 2013 and the results are summarized in Figure 5.4. Each point is connected with a straight line to visualize the general evolution and variability of the feed water.

It can be noticed from the feed water characteristics that phosphate and ATP values are very low and almost at the limit of detection of the equipment. Phosphate concentration is below the detection limit in some analyses and is represented as zero in the plot; however, due to the difficulty to measure phosphate at low levels, it cannot be confirmed that it is absent [10].

Overall water composition remains reasonably constant during the whole year and shows a relatively low biofouling potential. TOC levels are low, at around 1.25 mg/L, regardless of the season; however, TDS and nitrate appear to slightly increase after summertime, around September.

5.2.6 Nutrients

Carbon, nitrogen, and phosphate are needed by bacteria to grow, reproduce, and eventually build a biofilm [11]. In order to promote biological fouling and thus reduce the duration of each experiment, readily bioassimilable nutrients are dosed in the feed water. A concentrated sodium salts solution of acetate, nitrate, and orthophos-

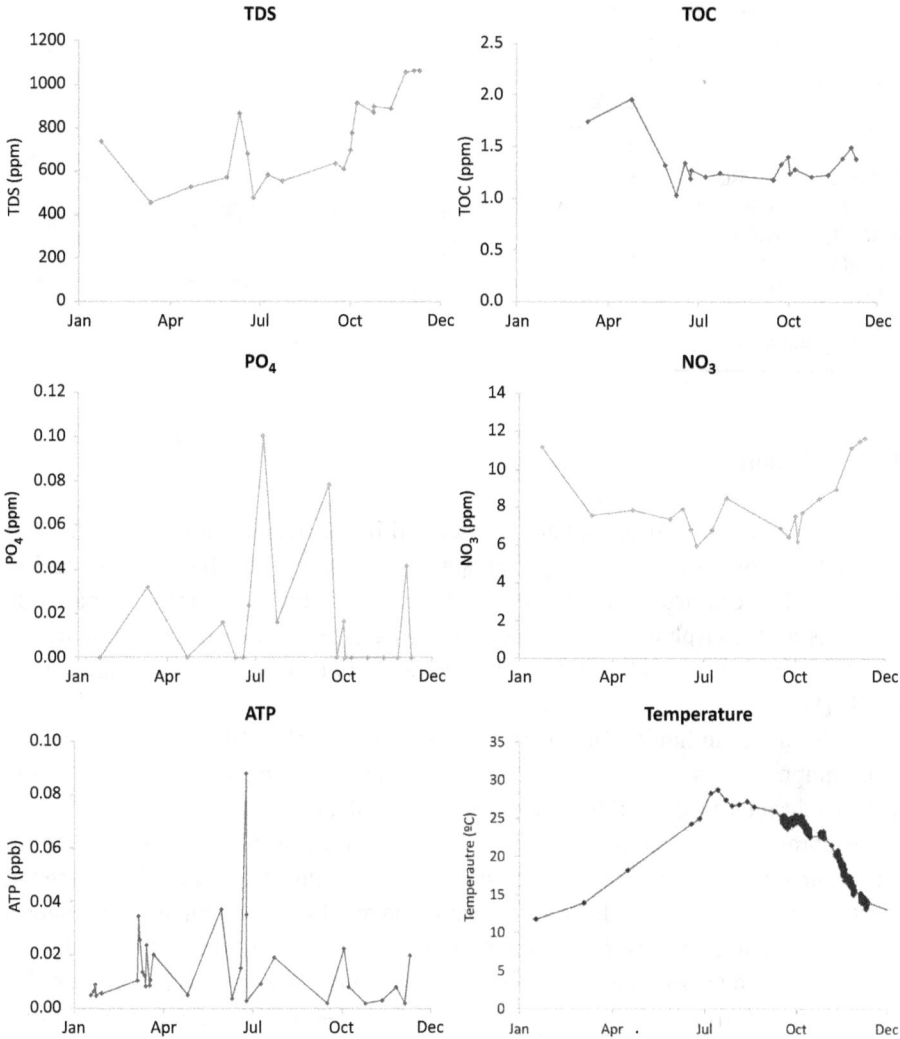

Figure 5.4: Feed water composition variation.

phate is prepared in a separated tank and added to the feed water using a peristaltic pump. This is done in order to supply enough food for bacteria to grow [11].

According to the typical ratios found in biomass and ensuring that phosphate is not restricting biofouling development [12], the ratio of C:N:P is set at 100:20:10. As inorganic salts can be directly used by bacteria, the concentration needed for accelerating biofouling is very low.

The calculation for injection rate is based on the feed water flow rate, the dosing pump frequency, and the tank concentration. The nutrient dosing calculations for one experiment is included in Table 5.4.

Table 5.4: Example for a dosing of 0.2 ppm of C in the feed water.

Parameter	Value
Feed flow	700 L/h
Nutrient pump stroke	80%
Nutrients dosing Pump	0.792 L/h
Nutrient tank autonomy	3.2 days
Dosing tank volume	60 L
CH_3COONa in tank	36.3 g
$NaNO_3$ in tank	12.9 g
$NaH_2PO_4 \cdot H_2O$ in tank	5.3 g

5.2.7 ATP analysis

ATP is the nucleoside triphosphate found in all living cells including bacteria. This molecule is used as a quick energy transfer unit in many endothermic biochemical reactions. This characteristic is the reason for its correlation with active biomass [13]. ATP acts as a phosphate group donor, releasing energy when the phosphodiester bond is hydrolyzed to adenosine diphosphate (ADP) or adenosine monophosphate (AMP) [14].

ATP content in liquid samples is analyzed using a Celsis Advance Luminometer. This equipment has a limit of detection of 2 ppt and the sampling volume is 100 μL. The membrane coupon and feed spacer analysis is done by cutting a 4 × 4 cm square of membrane and spacer. After this step, it is submerged in 20 mL ultrapure water to extract and dissolve the biofilm. A physical removal of the attached biofilm is used to extract the fouling layer with 2 min sonication on an ultrasonic bath at room temperature. The liquid sample is transferred into a sterile Eppendorf, where it can be immediately analyzed or frozen to −20 °C (−4 °F) for maximum 7 days. This is done in order to prevent ATP from being damaged. Samples are then measured using calibration curves, converting light signal to ATP concentration. ATP results are expressed as ng/cm^2 according to eq. (5.2). Dilution factor herein considered is a 100 times factor. Storage and extraction effects are optimized in the laboratory to reduce analytical variability. It should be noted that 0.02 L corresponds to the extraction volume, and 16 cm^2 corresponds to the membrane surface.

ATC calculation example

$$0.8\text{ppb ATP} \cdot \frac{100}{1} \cdot \frac{0.02\,\text{L}}{16\,\text{cm}^2} \frac{1{,}000\,\text{ng}}{1\,\mu g} = \frac{100\,\text{ng}}{\text{cm}^2}\text{ATP} \tag{5.2}$$

5.2.8 TOC analysis

TOC content in liquid samples is analyzed using TOC-L Shimadzu using UNE EN-1484:1998 method. The sample is oxidized via high temperature catalytic combustion and quantified using an infrared detector. The quantification is performed using a calibration curve obtained from a TOC standard solution. The equipment has a limit of detection of 0.1 ppm and the sampling volume is 50 µL.

Membrane coupon and feed spacer TOC analysis is done using the same extraction procedure that is applied for ATP. Details are shown in Section 5.2.7. Samples are kept still for at least one hour after the extraction step, to allow bigger particles in the suspension to precipitate. This step is done in order to prevent particles from plugging the instrument and damaging it. It should be then noticed that dissolved organic carbon might be a more accurate definition of this measurement.

The sample is either directly analyzed or stored at 5 °C after sample acidification. This is done to prevent the organic compounds present in the sample from decomposing.

Samples are measured and expressed as TOC concentration using the equipment's internal calibration curve. Taking into account the size of the surface analyzed and the extraction volume, TOC results are expressed as mg/m^2. The calculation steps are done using TOC concentrations in eq. (5.2).

5.3 Results and discussion

In general, it is recognized that biofilms are formed as a defense mechanism by bacteria to protect them from their surroundings. Bacteria tend to generate biofilms when they are under stress, and they are thought to attach to surfaces that have a conditioning layer. Once they are attached, the bacteria propagate and/or build more biofilm by feeding off the trace organic contaminants in the water. If the water source no longer has nutrients, bacteria will survive by switching to a dormant state or feed off the polysaccharides in their biofilm.

This general understanding of biofilms is used to identify parameters to optimize and control, in an attempt to develop a one-week flat cell screening test. Feed channel shear is expected to impose stress on the bacteria and promote biofilm formation. The shear is related to the velocity of the water flowing through the feed channel and the resistance of the feed spacer. Higher permeate flux is expected to accumulate more organics on the membrane's surface and more quickly form the surface conditioning layer for biofilm initiation. The concentration of nutrients in the feed water and water temperature will also impact the rate of biofilm formation.

5.3.1 Initial biofouling protocol setup

Initial probing experiments identified the following conditions as providing good bio-film within 3 days: 28-mil feed spacer, a cross-flow velocity of 1.2 m/s (feed flow = 2 L/min), operating flux of 34 LMH, and a 0.4 ppm C feed water nutrient loading. The flux or rejection of the membranes showed little change over the course of the experiment, as can be observed in Figure 5.5, but the membrane and spacer ATP (205 ± 166 ng/m²) and TOC (122 ± 16 mg/m²) values clearly demonstrated effective biofilm growth.

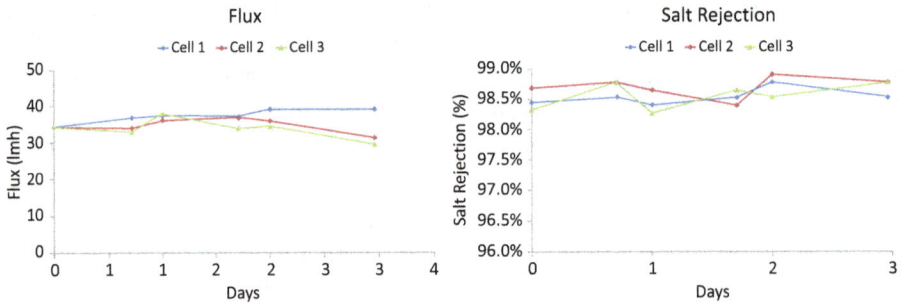

Figure 5.5: Operating results' example for flux and salt rejection.

These conditions are repeated; however, not only are the standard deviations of the measurements within each test large but also between tests, as shown in Table 5.5. Further optimization is completed to minimize the standard deviation.

Table 5.5: Initial probing experiment results.

C (ppm)	N (ppm)	P (ppm)	Flux (LMH)	Flow velocity (m/s)	ATP (ng/cm² ± 1σ)	TOC (mg/m² ± 1σ)	Flux loss (% ± 1σ)
0.4	0.08	0.04	34	1.2	205 ± 166	122 ± 15	3 ± 15
0.4	0.08	0.04	34	1.2	132 ± 51	233 ± 112	15 ± 4

5.3.2 Cross-flow velocity optimization

Since the cross-flow velocity is much higher than typically observed in an element, exploring the effect of lowering it is pursued. High cross-flow velocity is expected to impose a good deal of stress on the bacteria, but it also may be very disruptive to the growing biofilm and cause sloughing, which may create measurement variability. Keeping all other conditions the same but reducing the cross-flow velocity from 1.2 to 0.6 m/s, provided a reduction in the variability, as shown in Table 5.6.

Table 5.6: Results al lower cross-flow velocity.

C (ppm)	N (ppm)	P (ppm)	Flux (LMH)	Flow velocity (m/s)	ATP (ng/cm$^2 \pm 1\sigma$)	TOC (mg/m$^2 \pm 1\sigma$)	Flux loss (% $\pm 1\sigma$)
0.4	0.08	0.04	33	0.6	139 ± 3	132 ± 8	5 ± 4
0.4	0.08	0.04	32	0.6	200 ± 38	105 ± 15	6 ± 6

5.3.3 Nutrient loading optimization

The variability may be a result of generating a highly mature biofilm that not only is forming but also re-dispersing (sloughing). Reducing the nutrient level in the feed and reducing the flux to reduce the amount of organic material deposited on the membrane surface is explored to improve biofilm growth reproducibility. In this series of tests, the membrane flux is lowered to 20 LMH and the loading of three nutrients at 0.4, 0.2, and 0.1 ppm C are tested. The results are summarized in Table 5.7.

Table 5.7: Nutrient loading effect on lower flux experiments.

C (ppm)	N (ppm)	P (ppm)	Flux (LMH)	Flow velocity (m/s)	ATP (ng/cm$^2 \pm 1\sigma$)	TOC (mg/m$^2 \pm 1\sigma$)	Flux loss (% $\pm 1\sigma$)
0.4	0.08	0.04	20	0.6	280 ± 50	91 ± 11	3 ± 3
0.2	0.04	0.02	20	0.6	76 ± 26	59 ± 3	3 ± 2
0.1	0.02	0.01	20	0.6	21 ± 5	17 ± 5	3 ± 2

5.3.4 Warm temperature method validation

Operating at lower flux and nutrient loading levels of 0.1 or 0.2 ppm C provide acceptable results of the three flat cells that are operated in parallel. Final validation of these conditions is completed by conducting several replicates. Good reproducibility and acceptable standard deviations are observed within each run but variability between runs still exists, which may be due to uncontrolled changes in the natural water composition. The results are summarized in Table 5.8. With these optimized conditions, it is also noted that very little change in flux and rejection is observed.

5.3.5 Temperature effect study

After having achieved a good confidence in the conditions of the method, it came as a surprise when suddenly experiments are conducted and very little biofilm forms. This can be observed in Figure 5.6. A thorough review of the operating conditions and

Table 5.8: Validation of the quick biofouling test at two nutrient loadings.

C (ppm)	N (ppm)	P (ppm)	Flux (LMH)	Flow velocity (m/s)	ATP (ng/cm^2 ± 1σ)	TOC (mg/m^2 ± 1σ)	Flux loss (% ± 1σ)
0.2	0.04	0.02	20	0.6	59 ± 8	28 ± 2	0.3 ± 1
0.2	0.04	0.02	20	0.6	98 ± 9	47 ± 2	5 ± 3
0.2	0.04	0.02	20	0.6	52 ± 3	56 ± 4	2 ± 1
0.1	0.02	0.01	20	0.6	34 ± 11	96 ± 35	2 ± 1
0.1	0.02	0.01	20	0.6	18 ± 7	15 ± 2	2 ± 2
0.1	0.02	0.01	20	0.6	8 ± 4	14 ± 3	0.7 ± 0.5

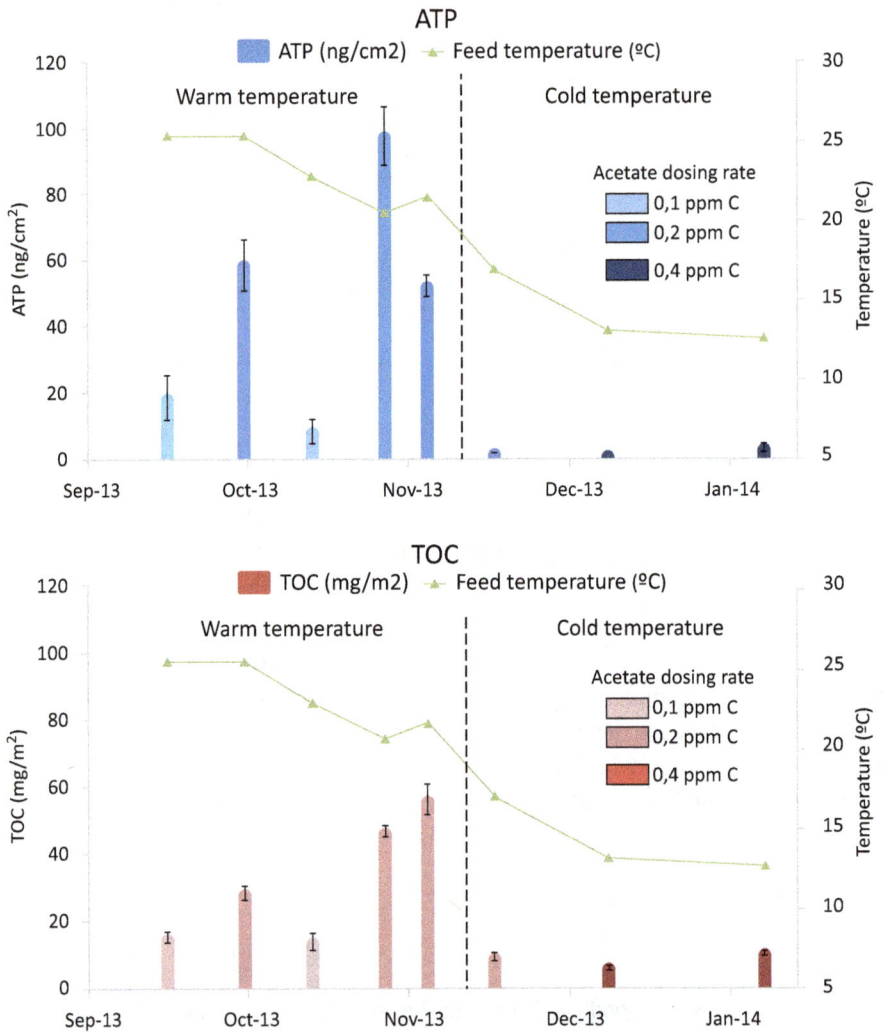

Figure 5.6: ATP and TOC evolution, considering seasonal effect.

nutrient dosing confirmed that no changes are made. However, the natural feed water temperature dropped below 20 °C due to seasonal changes. Since resources are not available to install a heat exchanger to control the feed water temperature, the biofouling method is adjusted to promote biofilm formation despite the low temperatures.

5.3.6 Cold temperature method validation

Increasing the shear stress on the bacteria by using only 22-mil feed spacers increases the duration of the experiment from 3 days to 5 days and increases the nutrient level up to 1.0 ppm C, provided there is acceptable biofilm formation, as measured by ATP and TOC. The results of two replicates experimental runs are summarized in Table 5.9.

Table 5.9: Summary of low-temperature experiments.

C (ppm)	N (ppm)	P (ppm)	Flux (LMH)	Flow velocity (m/s)	ATP (ng/cm$^2 \pm 1\sigma$)	TOC (mg/m$^2 \pm 1\sigma$)	Flux loss (% $\pm 1\sigma$)
1.0	0.2	0.1	20	0.6	44 ± 4	45 ± 7	3 ± 2
1.0	0.2	0.1	20	0.6	17 ± 11	48 ± 3	1 ± 1

The experimental throughput when the water temperatures are below 20 °C is reduced and limited to only studying 22-mil feed spacer geometries and membrane chemistry effects, but it allows fundamental research to continue nonetheless.

5.4 Conclusions

These experiments have validated the capability to evaluate biofouling on flat cell membranes in a controlled and reproducible way when no temperature control is available. The results obtained showed that despite seasonal temperature, an early-stage biofilm can be obtained in a short time period by dosing nutrients in the feed water.

Therefore, a method for water temperatures above 17 °C is obtained in order to overcome slower kinetics on biofilm growth when temperatures are lower. The summertime protocol consists of a 3-day test with a nutrients dosage of 0.2 mg/L carbon. The wintertime protocol consists of a 5-day test with a nutrients dosage of 1.0 mg/L carbon. Both methods show a standard deviation in accordance with other methods found in the bibliography.

In this sense, a high throughput tool is available to test different membrane chemistries and feed spacer configurations. This pilot unit will be also useful in the basic understanding of fouling behavior and development, as well as the factors affecting biofouling. Eventually, this quick method will shorten the product evaluation and development time, and will enhance competitiveness by accelerating the launch of novel fouling resistant elements.

References

[1]   H. C. Flemming. Reverse osmosis membrane biofouling. Experimental Thermal and Fluid Science, 14(4) (1997) 382–391.

[2]   S. R. Suwarno, X. Chen, T. H. Chong, V. L. Puspitasari, D. McDougald, Y. Cohen, S. A. Rice, A. G. Fane. The impact of flux and spacers on biofilm development on reverse osmosis membranes. Journal of Membrane Science, 405 (2012) 219–232.

[3]   J. Vrouwenvelder, J. Kruithof. Biofouling of Spiral Wound Membrane Systems, IWA Publishing. (2011)

[4]   L. Vanysacker, P. Declerck, I. Vankelecom. Development of a high throughput cross-flow filtration system for detailed investigation of fouling processes. Journal of Membrane Science, 442 (2013) 168–176.

[5]   C. M. Pang, P. Hong, H. Guo. Biofilm formation characteristics of bacterial isolates retrieved from a reverse osmosis membrane. Environmental Science & Technology, 39 (2005) 7541–7550.

[6]   W. Lee, C. H. Ahn, S. Kim. Evaluation of surface properties of reverse osmosis membranes on the initial biofouling stages under no filtration condition. Journal of Membrane Science, 351 (2010) 112–122.

[7]   T. H. Chong, F. S. Wong, A. Fane. The effect of imposed flux on biofouling in reverse osmosis: Role of concentration polarisation and biofilm enhanced osmotic pressure phenomena. Journal of Membrane Science, 325 (2008) 840–850.

[8]   D. J. Miller, P. A. Aráujo, P. B. Correia. Short-term adhesion and long-term biofouling testing of polydopamine and poly(ethylene glycol) surface modifications of membranes and feed spacers for biofouling control. Water Research, 46 (2012) 3737–3753.

[9]   S. S. Bucs, A. I. Radu, V. Lavric, J. S. Vrouwenvelder, C. Picioreanu. Effect of different commercial feed spacers on biofouling of reverse osmosis membrane systems: A numerical study. Desalination, 343 (2014) 26–37.

[10]   X. Zhu, J. Ma. Recent advances in the determination of phosphate in environmental water samples: Insights from practical perspectives. TRAC Trends in Analytical Chemistry, 127 (2020) 115908.

[11]   J. D. Jacobson, M. D. Kennedy, G. Amy, J. C. Schippers. Phosphate limitation in reverse osmosis: An option to control biofouling? Desalination and Water Treatment, 5 (2009) 198–206.

[12]   J. S. Vrouwenvelder, M. C. M. Van Loosdrecht. Phosphate limitation to control biofouling. Water Research, 44 (2010).

[13]   N. Oulahal-Lagsir, A. Martial-Gros, M. Bonneau, L. J. Blum. Ultrasonic methodology coupled to ATP bioluminescence for the non-invasive detection of fouling in food processing equipment-validation and application to a dairy factory. Journal of Applied Microbiology, 89 (2000) 433–441.

[14]   C. A. Reddy, T. J. Beveridge, J. A. Breznak, G. A. Marzluf, T. M. Schmidt, L. R. Snyder. Methods for General and Molecular Microbiology, 3rd edition, American Society for Microbiology (ASM). (2007).

[15]   V. D. Villanueva, J. Font, T. Schwartz, A. M. Romaní. Biofilm formation at warming temperature: Acceleration of microbial colonization and microbial interactive effects. Biofouling, 27 (2011) 59–71.

[16]   H. C. Flemming, J. Wingender, U. Szewzyk. Biofilm Highlights, Springer Science & Business Media, 5. (2011).

[17] D. Van der Kooij, H. R. Veenendaal. Biofilm formation and multiplication of Legionella in a model warm water system with pipes of copper, stainless steel and cross-linked polyethylene. Water Research, 39 (2005) 2789–2798.

[18] N. Oulahal-Lagsir, A. Martial-Gros, M. Bonneau. Ultrasonic methodology coupled to ATP bioluminescence for the non-invasive detection of fouling in food processing equipment – Validation and application to a dairy factory. Journal of Applied Microbiology, 89 (2000) 433–441.

[19] K. J. Varin, N. H. Lin, Y. Cohen. Biofouling and cleaning effectiveness of surface nanostructured reverse osmosis membranes. Journal of Membrane Science, 446 (2013) 472–481.

[20] DuPont™. FilmTec™ Reverse Osmosis Membranes Technical Manual, Form No. 609-00071-0808.

Chapter 6
Differentiating between biofouling and organic fouling during operation

Using low energy reverse osmosis membranes for secondary wastewater reclamation – A compari-
son of Membrane A and Membrane B chemistries

This experiment compares two different membranes – membrane A and membrane B. Both membranes are designed to have different physical properties. They are compared during an initial organic fouling phase, followed by a second biological fouling phase, with the goal to understand which of their design parameters make them more suitable to gain biofouling and organic fouling resistance. Both membranes are operated side-by-side in exact operating conditions. In these two phases, it can be seen how membrane B allows for an improved fouling resistance as it only lost 23% permeability, while during the same period of time and in the same operating conditions, membrane A lost 46% permeability. This might indicate that a membrane with a smaller contact angle, therefore indicating a more hydrophilic membrane, might help in preventing fouling accumulation on the membrane. Moreover, these results might indicate that a membrane with a higher zeta potential, therefore indicating a stronger electrostatic repulsion between particles or foulants and the membrane, can lead to better dispersion and reduction of fouling.

6.1 Introduction

As has been described in the introduction, in natural waters, there are two natural phenomena.

The first one is typically characterized by an organic fouling period, where organics quickly absorb on the membrane surface, leading to a quick decrease in water permeability, as well as a typical improvement in the salt rejection [1]. This usually happens because as the membrane gets fouled, it is more difficult for water to pass through it, but it is also difficult for salt to pass through the membrane. During this period, pressure drops tend to be constant, as there is no biofilm blocking its feed-concentrate channel.

Acknowledgments: The author would like to acknowledge the contributors of this chapter: Gerard Massons, Guillem Gilabert-Oriol, Jon Johnson, Steve Jons, and Tina Arrowood. The authors would like to acknowledge all the members of the DuPont Water Solutions team for their support in this research.

This chapter is in the process of being submitted for publication, and was submitted to the International Desalination Association (IDA) conference in San Diego (USA) in September 2015.

https://doi.org/10.1515/9783111639079-006

The second phase is typically characterized by biological fouling, which develops slowly over time, and needs a bit more time to begin to be noticed. It is typically noticed by an exponential pressure drop growth over time. This is the consequence of a biofilm being formed, which blocks the feed-concentrate water flow, leading to a pressure drop increase over time [2].

6.2 Materials and methods

6.2.1 Wastewater characterization

Vila-seca wastewater treatment plant (WWTP) treats municipal wastewater from Vila-seca, La Pineda, and Salou villages. The intake of the water is from the effluent of the clarifier, which is then pretreated using two DuPont™ Ultrafiltration SFP-2880 XP modules. The average ionic and organic characteristics of the ultrafiltered water used for the reverse osmosis experiments are summarized in Table 6.1.

Table 6.1: Wastewater feed characterization.

Parameter	Average concentration
ATP	0.03 ± 0.02 ng/L
TOC	8 ± 1.5 mg/L
Calcium	135 ± 16 mg/L
Chloride	509 ± 48 mg/L
Nitrate	63 ± 26 mg/L
Sodium	308 ± 30 mg/L
Sulfate	290 ± 49 mg/L
Conductivity	2,408 ± 442 µS/cm
pH	7.3 ± 0.2

6.2.2 Experimental plant

The different trials were performed in the Fouling Resistant Elements Pilot Plant (FREPP), located in the Vila-seca WWTP. A single high-pressure centrifugal pump is used to feed the eight RO modules as shown in Figure 6.1. To match flux and recovery of elements that have different permeability, each pressure vessel has a needle valve in the permeate port to apply individual backpressure to each element. In the first trials, backpressure was registered manually using analog pressure indicators. On a later phase, automatic pressure transmitters were installed with the intention of tracking the changes on backpressure on a continuous mode. The plant can also operate on recirculation mode using activated carbon filters, which are used during stabilization phase in order to capture any organics that the elements might leach to the feed water. This experiment was performed in the Global Water Technology Center that DuPont Water Solutions has in Tarragona, Catalonia, Spain.

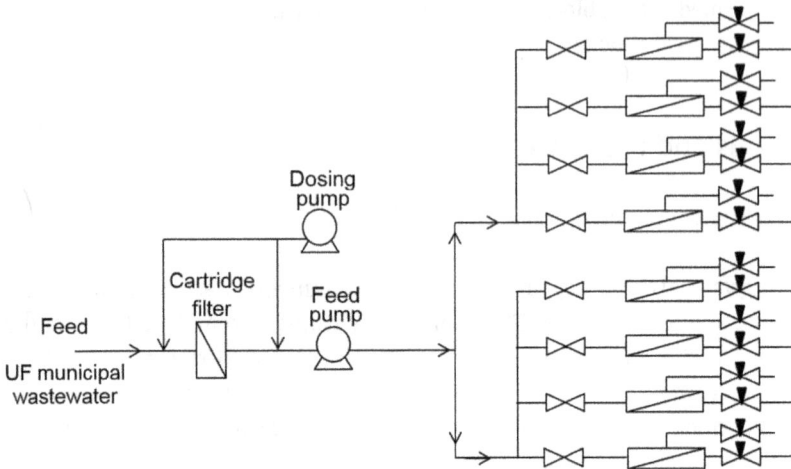

Figure 6.1: Fouling Resistant Elements Pilot Plant scheme.

6.2.3 Membranes used

Smoother, more hydrophilic, and lower surface-charged membranes are described in the literature as being more fouling-resistant [3–6]. Therefore, two 1812 membrane elements, membrane A and membrane B were chosen in this study to represent examples of a low and high fouling-resistant membrane surface, respectively. The properties of the two membranes are summarized in Table 6.2. Membranes A and B have similar degrees of roughness, but membrane A is more hydrophobic and more negatively charged than membrane B and is expected to be fouled more rapidly. The values for the atomic force microscopy (AFM), the contact angle, and the zeta potential are displayed in the table below.

Table 6.2: Membrane characteristics.

Parameter	Membrane A	Membrane B
AFM average roughness	61 nm	61 nm
Contact angle	72°	43°
Zeta potential (pH 8)	−25 mV	−5 mV

6.2.4 Experimental setup

Both 1812 membrane elements where first stabilized through a 5-min flushing using reverse osmosis permeate water with a salinity of 1,000 ppm NaCl and 26 LMH of flux. This flushing was discarded and sent to drain. Then the stabilization protocol followed by operating during 2 days in recirculation mode using a solution made of reverse osmosis permeate with 1,000 ppm NaCl and a flux of 26 LMH. During this recirculation, water was going through an activated carbon filter, so that any organics that might be present in the water being recirculated can be trapped into the activated carbon filter and any organic that might be present being recirculated back to the membrane feed can be avoided.

After this stabilization period, membranes were operated at a constant flux of 18 LMH and at a constant recovery of 5%.

6.3 Results and discussion

6.3.1 Pressure drop

The pressure drop of both elements, plotted in Figure 6.2, remained the same, as both had the same feed spacer. The step changes in pressure drop correspond to readjustments to maintain feed flow constant. As no biocide was dosed, a relatively rapid pressure drop increase was observed over the 35 days of operation due to natural biofouling development.

It should be noted that in this experiment, the first period, called the organic fouling period can be clearly seen. This period is characterized by organic fouling rapidly depositing on the membrane surface, leading to quick initial water permeability (A-value) loss, while pressure drop stays flat, as there is still no biofouling developing on the membrane surface.

The second period starts when biofouling starts to increase. This is defined as the biofouling period, and it is typically characterized by a noticeable pressure drop increase over time. This is caused by the bacteria that start to grow a biofilm, causing an obstruction of the water flowing through the feed-concentrate channel of a reverse osmosis membrane. It should be noticed that typically this biofouling period starts when a membrane is already organically fouled.

6.3.2 Water permeability

This test was run at relatively high flux rates (26 L/m^2 h) and without biocide dosing. Although the manual recording of backpressure reading increased the scattering on permeability results, the trends of both membrane chemistries can be observed in Figure 6.3.

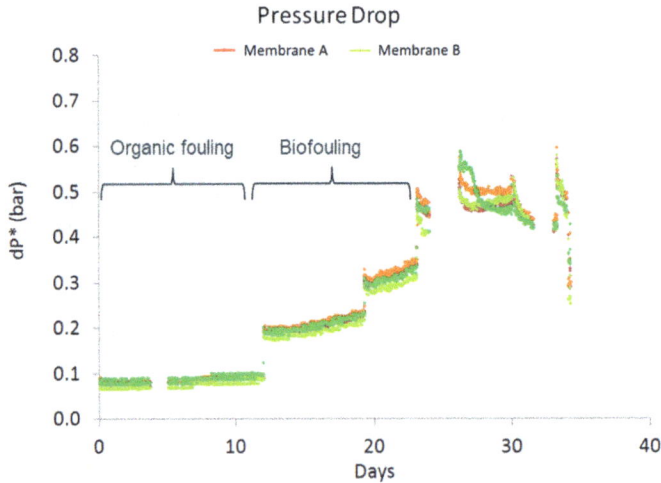

Figure 6.2: Pressure drop for membrane A and membrane B.

Membrane A always showed a higher A-value than membrane B, thus saving energy, despite a higher A-value loss over time. It should be noted that membrane B provided improved fouling resistance, as it only lost 23% permeability, while during the same period of time and in the same operating conditions, membrane A lost 46% permeability.

In this plot the initial flat tendency of the water permeability (A-value) evolution during the first two days can be seen. This is due to the stabilization procedure. Once the trial starts after the second day, a sharp decrease in water permeability can be observed. This is due to the organics quickly absorbing on the membrane surface.

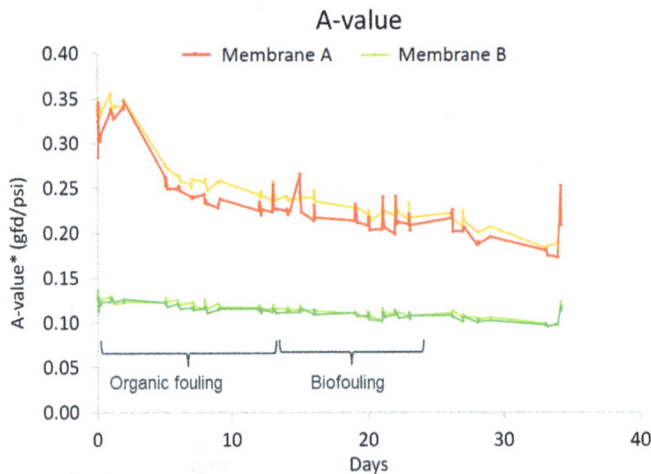

Figure 6.3: Water permeability for membrane A and membrane B.

6.3.3 Salt passage

A clear differentiation could be observed in salt passage, as shown in Figure 6.4. As expected, membrane A showed higher water permeability and salt permeance than membrane B. It should be noticed that salt permeance (B-value) quickly decreases during the initial organic fouling period. This is because as the membrane gets fouled with organic fouling, its active layer becomes thicker, thus making it more difficult for salts to pass through the membrane.

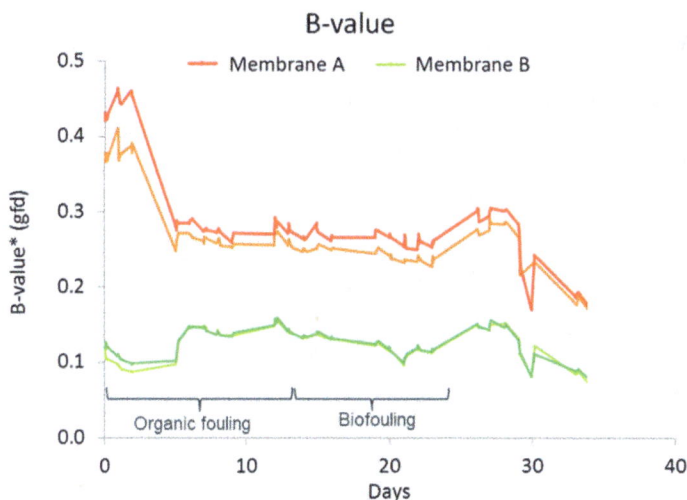

Figure 6.4: Salt permeance for membrane A and membrane B.

6.4 Conclusions

This experiment shows an initial organic fouling phase, followed by a second biological fouling phase. In these two phases, it can be seen how membrane B allows for an improved fouling resistance as it only lost 23% permeability, while during the same period of time and in the same operating conditions, membrane A lost 46% permeability. This might indicate that a membrane with a smaller contact angle, therefore indicating a more hydrophilic membrane, might help in preventing fouling accumulation on the membrane. Moreover, these results might indicate that a membrane with a higher zeta potential, therefore indicating a stronger electrostatic repulsion between particles or foulants and the membrane, can lead to better dispersion and reduction of fouling.

References

[1]   M. Xie, J. Lee, L. D. Nghiem, M. Elimelech. Role of pressure in organic fouling in forward osmosis and reverse osmosis. Journal of Membrane Science, 493 (2015) 748–754.

[2]   J. S. Vrouwenvelder, D. G. Von der Schulenburg, J. C. Kruithof, M. L. Johns, M. C. M. Loosdrecht. Biofouling of spiral-wound nanofiltration and reverse osmosis membranes: A feed spacer problem. Water Research, 43(3) (2009) 583–594.

[3]   M. Herzberg, M. Elimelech. Biofouling of reverse osmosis membranes: Role of biofilm enhanced osmotic pressure. Journal of Membrane Sciences, 295 (2007) 11–20.

[4]   M. Herzberg, S. Kang, M. Elimelech. Role of extracellular polymeric substances (EPS) in biofouling of reverse osmosis membranes. Environmental Science & Technology, 43 (2009) 4393–4398.

[5]   E. M. Van Wagner, A. C. Sagle, M. M. Sharma, Y. H. La, B. D. Freeman. Surface modification of commercial polyamide desalination membranes using poly (ethylene glycol) diglycidyl ether to enhance membrane fouling resistance. Journal of Membrane Science, 367 (2011) 273–287.

[6]   E. Matthiasson. The role of macromolecular adsorption in fouling of ultrafiltration membrane. Journal of Membrane Science, 16 (1983) 23–36.

Chapter 7
Differentiating between biofouling and organic fouling on a membrane surface

Method for distinguishing between abiotic organic and biological fouling of reverse osmosis elements used to treat wastewater

Fouling is one of the issues hindering the long-term performance of reverse osmosis (RO) systems. The aim of this study is to selectively quantify the extracellular polymeric substances (EPSs) from the organic substances present in the samples of autopsied elements. Samples were collected from elements used to treat wastewater, under two different fouling scenarios. In the first, fouling was mainly organic, while in the second, biofouling was promoted by dosing nutrients. The overall percentage of carbohydrates and proteins as organic foulants was quantified. In the first test, the percentage of the foulant consisting of EPS was only 19–34% versus approximately 100% in the biofouling scenario. At the same time, the increase in the differential pressure between the elements after the first test was only 50%, but more than 500% after the second test. This confirms that quantifying the EPS percentage of the foulant can be useful for determining the extent of biofouling versus abiotic organic fouling. Finally, it can be observed how biofouling development is strongly dependent on the availability of bioassimilable nutrients.

7.1 Introduction

7.1.1 Fouling in reverse osmosis

RO is often used as one of the most cost-effective strategies for producing high-quality water for a variety of applications. However, fouling is still one of the major hurdles in membrane technology, especially in RO systems, because it increases the energy needed

Acknowledgments: The author would like to acknowledge the contributors of this chapter: Gerard Massons-Gassol, Guillem Gilabert-Oriol, Veronica Gomez, Ricard Garcia-Valls, Veronica Garcia Molina, and Tina Arrowood. The author would also like to thank all teams of DuPont Water Solutions for their support in doing this research and specifically, Javier Dewisme and Javier de la Fuente for their reliable work in operating the pilot plant at Vilaseca WWTP and Mireia Font, and Patricia Carmona for the excellent analytical support.

This chapter was originally published with the following references: G. Massons-Gassol, G. Gilabert-Oriol, V. Gomez, R. Garcia-Valls, V. Garcia-Molina, and T. Arrowood Method for distinguishing between abiotic organic and biological fouling of reverse osmosis elements used to treat wastewater. Desalination and Water Treatment, 83 (2017) 1–6.

https://doi.org/10.1515/9783111639079-007

and also requires frequent shutdowns for cleanings [1]. When designing a new system, it is very complex to predict the severity of fouling that might result in a higher frequency of cleaning and reduced productivity. Therefore, improving the ability to predict, troubleshoot, and reduce the fouling of RO systems continues to be a topic of great interest.

There are different types of membrane fouling. The two most problematic types are biological and organic [2]. Biological fouling is usually associated with an increase in differential pressure in the first-stage pressure vessel [3–5]. Organic fouling usually causes an increase in resistance to transport water through the membrane, which reduces permeate flow [6]. Both types of fouling usually occur together, so it is difficult to optimize systems, because steps to improve biological fouling may worsen organic fouling and vice versa. The characteristics and distribution of each type of fouling must be studied if membrane performance is to be more sustainable [7]. Determining whether biological or organic fouling is dominant is important for designing improvements in pretreatment. If biopolymers such as proteins or carbohydrates are identified as the primary type of fouling, biocide dosing or nutrients limitation using biological pretreatment would likely improve RO fouling problems [8]. However, if the source of the foulants is not determined and fouling was actually caused by abiotic organic compounds, these same pretreatments solutions might result in inefficient fouling control. Unfortunately, commonly used fouling quantification techniques do not reveal the source of the organic compounds present in fouling samples. Information about the type of RO contamination is also needed so that cleaning recommendations can be adapted to the pilot unit systems [9].

7.1.2 Fouling characterization

System fouling was assessed by monitoring the performance and analytically characterizing the membrane. Changes in permeate flow rate, feed pressure, salt passage, and pressure drop over time were symptoms of the onset and severity of fouling. Autopsying the elements after operation and analyzing the composition of the foulant indicates the source of the foulant. The measurement of adenosine triphosphate (ATP), total organic carbon (TOC), and total nitrogen (TN) are the most common analytical characterization techniques used to study both organic and biological fouling [10]. In most cases, biological and organic fouling are found together [11–12]. Only system differential pressure and foulant ATP concentration are correlated with biological fouling [13].

7.1.3 Biofouling vs. organic fouling: characteristics and composition

The main component of biofouling is a polymer matrix extracted by bacteria [14]. This matrix is a strongly hydrated mixture of polysaccharides, proteins, nucleic acids, and lipids, known as EPSs [5, 15]. Polysaccharides and proteins are, on a mass basis, the main components of the biofilm matrix [16]. The amount and type of protein and carbohydrate found in EPS depends on the bacteria strain, environmental conditions, and stress events [17–18]. Carbohydrates and proteins are rich in organic carbon, and peptides are also rich in nitrogen, so TOC and TN concentrations are high when biofilms are analyzed [19].

Organic fouling occurs when organic compounds found in the feed water are deposited on the membrane surface. These compounds usually contain carbon and nitrogen, and, therefore, can be detected by a positive response for TOC and TN.

It has been suggested that comparing ATP and TOC levels in a foulant is one way of determining whether organic foulants come from biofilm formation or abiotic compounds [1]. However, ATP degrades quickly and is highly sensitive to external factors [20] that can influence the bacterial metabolic state (chemical cleaning, biocides, etc.), so ATP levels can be unreliable. The ATP concentration does not correlate with the presence of EPS under certain conditions [21–22]. The presence of biomass, rather than its physiological activity, is usually the root cause of fouling problems in RO systems [23]. Nonetheless, TOC and TN levels alone did not differentiate the source of the foulants quantified (biological or abiotic).

Liquid chromatography–organic carbon detection–organic nitrogen detection (LC-OCD-OND) has emerged as a useful technique for identifying and quantifying the various fractions of the natural organic matter pool (protein and polysaccharide, humic, fulvic, building blocks, and low-molecular-weight organics) [24]. This technique has been successfully used in some studies to calculate the fraction of organic carbon associated with biopolymer (proteins and polysaccharides) [25–26]. However, the results can be complex to interpret, and the technique is generally not available for routine membrane fouling samples [27]. Other studies suggest that the proportion of biopolymer on a membrane foulant sample can be calculated using the area of the pyrochromatograms, obtained using pyrolysis gas chromatography – mass spectrometry [28]. Like the LC-OCD-OND method, analyzing samples is time consuming and technically challenging.

This chapter reports a simplified approach to determining the relative extent of biofouling over organic fouling in RO samples. Carbohydrates and proteins are the main constituents of the biofilm matrix, and whether they are present or not, can discern between biological and organic foulants [29]. The proportion of carbohydrates and proteins in the TOC and TN pool will be calculated to determine the fraction of biological and abiotic carbon and nitrogen, respectively. To validate the viability of the approach, the protocol will be applied to determine the EPS fraction of various RO samples operating under two different conditions. The protocol uses techniques that are available at most analytical laboratories. The proportions intend to provide information about the source and proportion of the compounds in complex fouling samples.

7.2 Materials and methods

RO elements were exposed to two different fouling conditions and autopsied to provide samples for analysis. The concentrations of the parameters quantified were compared to correlate different biopolymer levels, according to the testing conditions. This experiment was performed at the Global Water Technology Center that DuPont Water Solutions has in Tarragona, Catalonia, Spain.

7.2.1 Wastewater characterization

Two separate RO element exposure tests were conducted, one with a high and one with a low biofouling tendency. Both tests used water from the secondary effluents collected from a full-scale wastewater treatment plant in Vila-seca, Catalonia, Spain. The typical composition of the inlet water can be seen in Table 7.1. The high chemical oxygen demand (COD) and TOC concentrations present an inherent organic fouling potential for the RO elements, mainly nonbiodegradable organic compounds, according to the ratio of COD to the biological oxygen demand (BOD5) [1]. To promote high biofouling levels, an external dosing pump was used to dose nutrients in the feed stream and to stimulate bacteria present in the feed water (trial with nutrients added) [30]. In the experiment aiming for a low biofouling tendency, no dosing was used (trial with no nutrients added).

Table 7.1: Characterization of feed wastewater.

Feed water	Concentration
Total dissolved solids (mg/L)	1,880
Total suspended solids (mg/L)	0.14
COD (mg/L O_2)	21.2
BOD5 (mg/L O_2)	2.1
TOC (mg/L)	6.0
ATP (ng/L)	38

7.2.2 Experimental plant

Different trials were performed at the Fouling Resistant Elements Pilot Plant (FREPP), located in the Vila-seca wastewater treatment plant (WWTP). A single high-pressure centrifugal pump is used to feed the eight RO modules, as shown in Figure 7.1. To match the flux and recovery of elements having different permeability, each pressure vessel has a needle valve in the permeate port to apply individual backpressure to each element. In the first trials, backpressure was registered manually using analog pressure indicators. On a later phase, automatic pressure transmitters were installed with the intention to track the changes on backpressure on a continuous mode. The plant can also operate on recirculation mode using activated carbon filters, which are used during the stabilization phase in order to capture any organics that the elements might leach to the feed water. This experiment was performed at the Global Water Technology Center that DuPont Water Solutions has in Tarragona, Catalonia, Spain.

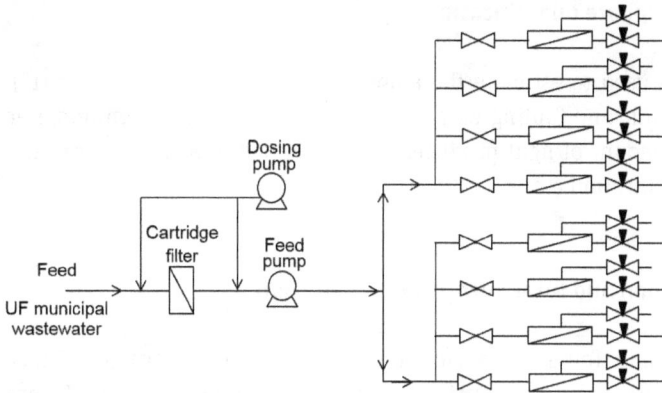

Figure 7.1: Fouling Resistant Elements Pilot Plant scheme.

7.2.3 Experimental setup

In each trial, either six (trial with nutrients added) or eight (trial with no nutrients added) 1.8-inch diameter by 12-inch long RO elements were assembled in parallel and allowed to treat the wastewater without any recycling. All elements were operated under similar conditions in both trials (10 bar, 4.5% recovery, and 25 L/m^{2}h) for approximately one week (temperature from 17 to 26 °C). To influence the type of fouling accumulated, nutrients were dosed in the feed water (trial with nutrients added). These include a source of carbon (0.1 mg/L C as acetate), nitrogen (0.02 mg/L N as nitrate), and phosphorous (0.01 mg/L P as phosphate). Even though they are present in lower levels than in the feed water, these compounds are readily bioavailable and promote rapid biofilm growth. After each test, the exposed elements were autopsied, and samples taken for analysis.

7.2.4 Membrane foulant extraction

After opening the elements lengthwise, a 4 × 4 cm (16 cm^2) sample from the middle region of the membrane and spacer was placed in a glass vial. A 20 mL phosphate-buffered saline (PBS, VWR) solution was added to dissolve the foulant present [31].

Strong acid cation resin (DOWEX® MARATHON™ C Na⁺, Dow Chemical) was added (1 g) to improve EPS solubility [32–33]. The vial was sonicated using an ultrasonic cleaning bath (FB15061, Fisher Scientific) at room temperature for 2 min in triplicate [34–35]. After this treatment, all the foulant was fully dissolved. Samples were stored at −21 °C until analysis.

7.2.5 Adenosine triphosphate quantification

ATP concentration was used to estimate the amount of viable biomass present [17]. ATP content in the membrane fouling extract was measured using a luminometer (Celsis Advance). The amount of light produced was converted to ATP concentration using the equipment calibration curve.

7.2.6 Total organic carbon and total nitrogen quantification

Both biological and organic foulants are rich in organic carbon and nitrogen. Thus, TOC and TN were good methods for capturing both types of fouling. TOC and TN were determined by a catalytic combustion using a TOC/TN analyzer (TOC-L Shimadzu), calibrated using potassium hydrogen phthalate and urea-BSA (1:1), respectively.

TN is the sum of total inorganic nitrogen (TIN) and total organic nitrogen (TON) in a sample. However, fouling analysis from previous studies, using the same wastewater, has shown that the nitrogen present in the foulant was over 92% organic [11]. Consequently, in this study, it was assumed that the inorganic nitrogen portion (NH_4, NO_3, and NO_2) was negligible, in comparison to the organic nitrogen. The TN analysis of the autopsied elements was assumed to account mainly for TON.

7.2.7 Carbohydrate quantification

The polysaccharides from EPS were measured using the Dubois method, also known as the phenol-sulfuric acid method [36]. The Dubois method has been widely reported for EPS polysaccharide quantification as a simple colorimetric method [37–40]. The carbohydrate concentration of the membrane fouling extract was detected colorimetrically using the Hach DR 5000 spectrophotometer ($\lambda = 490$ nm).

Glucose (Sigma Aldrich) was used to calibrate the method. Glucose molecules (180 g/mol) contain a significant amount of carbon (40 wt% C). This factor was used to convert glucose concentration to carbohydrate carbon ($C_{carb.}$) [25], so that it could be compared with the TOC measurements.

7.2.8 Protein quantification

Bicinchoninic acid method (BCA) [41, 42] was selected to quantify proteins in the membrane-fouling extract. The BCA method can be readily used as a fast and simple colorimetric kit (Micro BCA™ Protein Assay Kit, Thermo Fisher). Absorbance was measured using the Hach DR 5,000 spectrophotometer ($\lambda = 562$ nm).

The bovine serum albumin (BSA, Sigma Aldrich) was used to calibrate the method. Albumin molecule (66,463 g/mol) contains a significant amount of nitrogen (16 wt% N) [43]. This factor was used to convert the BSA concentration to protein nitrogen ($N_{proteins}$), so that it could be compared with the TN measurements.

7.2.9 EPS fraction quantification

To compare the carbohydrate and protein results with the TOC and TN, the responses of calibration compounds on the TOC-L analyzer (Shimadzu) were checked.

For the comparison of the organic carbon, the glucose TOC results are the same as the theoretical carbon percentage (40 wt% C). The TOC results for BSA showed that it contained a 2.5 wt% C, a factor that was later used to calculate the protein carbon ($C_{protein}$). This percentage was lower than the reported BSA elemental composition [43]; it might be due to a low oxidation yield of the BSA carbon.

The sum of the protein and carbohydrate carbon divided by the TOC result gave the theoretical fraction of organic carbon associated with the EPSs, according to eq. (7.1).

EPS in TOC

$$\text{EPS in TOC (\%)} = \frac{C_{carb.} + C_{protein}}{TOC} \cdot 100 \tag{7.1}$$

Likewise, the proportions of nitrogen in BSA and glucose solutions were determined using the TN method. As expected, glucose had no response, as no nitrogen is present in the molecule. The nitrogen proportion measured for BSA was in agreement with its nitrogen composition (16 wt% N).

The percentage of protein nitrogen divided by the TN result expressed the fraction of organic nitrogen associated with EPS, according to eq. (7.2).

EPS in TN

$$\text{EPS in TN (\%)} = \frac{N_{protein}}{TN} \cdot 100 \tag{7.2}$$

Potential interferences with the BCA and Dubois method measurements were also discarded. The presence of glucose and BSA in the sample did not affect the quantification of proteins and carbohydrates, respectively.

7.3 Results and discussion

ATP, TOC, TN, $C_{carb.}$, $C_{protein}$, and $N_{protein}$ were measured in samples of the elements exposed to the trial referred as "with nutrients added" conditions or the trial referred as "with no nutrients added" environment.

7.3.1 Wastewater trial with nutrients added

The measured analytical parameters and the corresponding calculations of the composition of the foulants from the elements of the trial "with nutrients added" are summarized in Figures 7.2 and 7.3. The biological fraction (C-carb., C-protein, and N-protein) almost matches the overall organic carbon and nitrogen measured. The material balance was not perfect because of accumulative errors of the various quantification methods involved. However, the results leave little doubt that when the nutrients were dosed, most of the foulants present on the membrane were biopolymers (carbohydrates and proteins).

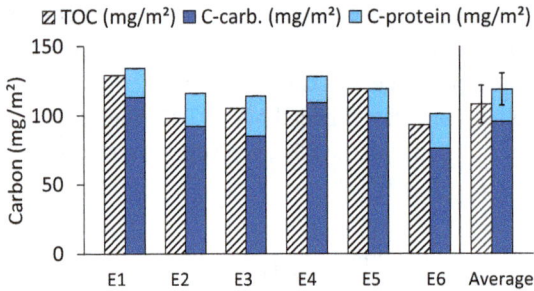

Figure 7.2: TOC and carbon from carbohydrates and proteins measurements from elements, E1-E6, operated with nutrients added to the feed water.

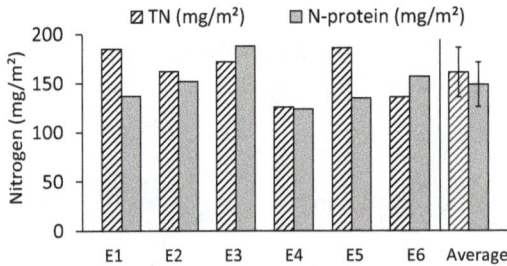

Figure 7.3: TN and nitrogen from proteins measurements from elements, E1-E6, operated with nutrients added to the feed water.

7.3.2 Wastewater trial with no nutrients added

In the conditions "with nutrients added," the carbohydrate and protein concentrations account for approximately all the TOC and TN present on the membrane. However, when no nutrients were dosed, carbohydrates and proteins account only for approximately one-fourth of the TOC or TN of the foulants (see Figures 7.4 and 7.5). The difference between the TOC and the carbon from carbohydrates and proteins, likely reflects fouling associated with the abiotic organic material present in the feed water as its high COD/BOD5 ratio of larger than 10 suggests [1].

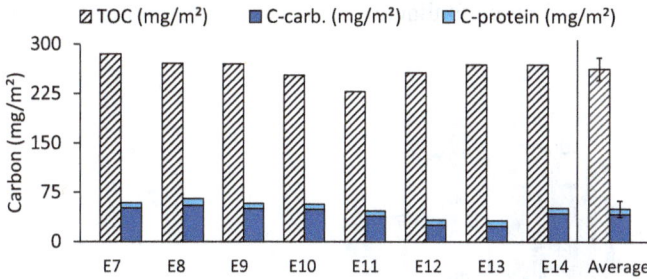

Figure 7.4: TOC and carbon from carbohydrates and proteins measurements from samples of elements, E7-E14, operated without nutrients added to the feed water.

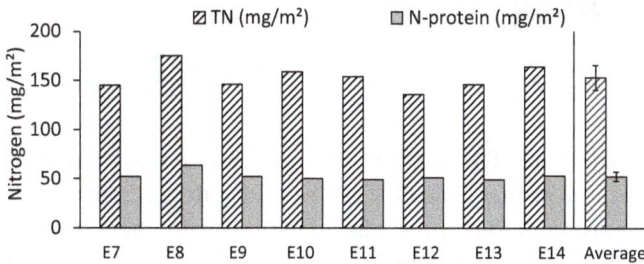

Figure 7.5: TN and nitrogen from proteins measurements from elements, E7-E14, operated without nutrients added to the feed water.

7.3.3 Correlation between EPS fraction and membrane performance

Using the equations described in Section 7.2.9, the EPS fraction was calculated for the elements from the test, with and without nutrients (Sections 7.3.1 and 7.3.2, respectively). The percentage of the measured TOC and TN attributed to carbohydrate and protein was a simple way of comparing the results of the two tests.

The different EPS fractions (based on organic carbon and nitrogen distribution) are plotted in Figure 7.6. The same graph also shows the percent increase in the measured feed concentrate pressure drop for each element at the end of the test. As expected, a clear correlation between the calculated EPS fraction and the dP increase can be observed. When nutrients were dosed (biofouling promoted), the pressure drop increased considerably and biopolymers accounted for almost all the organic nitrogen and carbon measured. However, in the samples from the "trial with no nutrients added," the differential pressure increases, and the EPS fractions were much lower. The method provided similar conclusions as other publications using LC-OCD-OND, where the biopolymer peak for samples containing biofouling was significantly larger than for samples containing organic fouling [44]. However, a clear correlation between carbohydrates and proteins, detected by photometric methods, and performance decline, caused by biofouling, was not observed in other publications [45].

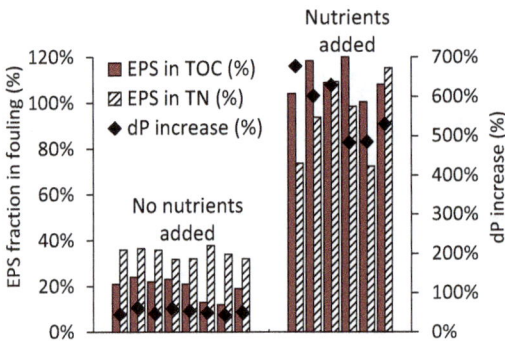

Figure 7.6: Correlation of membrane performance and EPS fraction for the two testing conditions.

These results can be summarized into Figure 7.7, where it can be seen that a biofilm is mainly composed of EPS, as well as that biofouling can be distinguished from organic fouling by looking at the portion of EPS in the whole fouling. Again, these pictures show the EPS as expressed in carbon molecules, as well as the EPS, as expressed in nitrogen molecules. Biofouling is characterized by having a high pressure drop, as can be seen from the plot.

A comparison of the ATP results reveals good correlation with EPSs percentages obtained in the two sets of samples. This data can be observed in Table 7.2. The operational and analytical results both showed that when nutrients were used, the fouling observed was mainly attributed to biofilm growth. When nutrients were dosed during the fouling test, the foulants found on the samples had higher concentrations of ATP as compared to foulants found on the elements operated without nutrients added. This highlights the importance of readily bioavailable nutrients to enhance bacteria colonization, reproduction, and biofilm formation.

EPS (in TOC)

Pressure Drop

EPS (in TN)

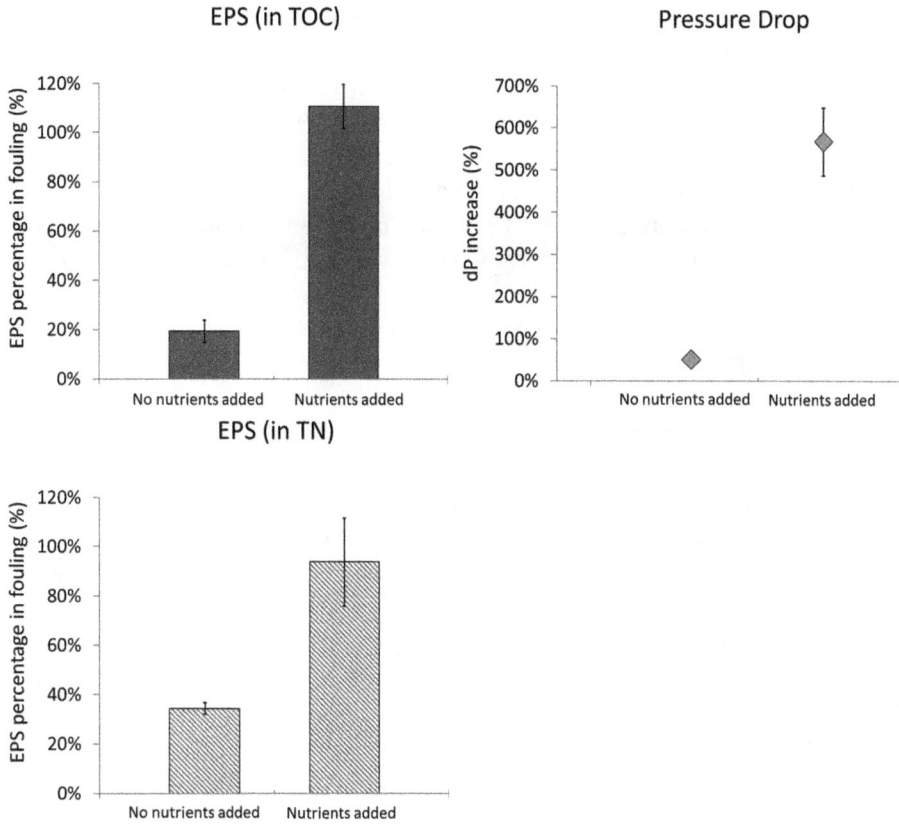

Figure 7.7: EPS and pressure drop increase relationship.

Table 7.2: Average ATP, dP, and EPS fraction summary for the two testing conditions.

Parameter	No nutrients added test	Nutrients added test
EPS in TOC (%)	19 ± 4	111 ± 9
EPS in TN (%)	34 ± 2	94 ± 18
ATP (ng/cm^2)	1 ± 0.1	40 ± 10
dP increase (%)	51 ± 5	568 ± 80

7.4 Conclusions

The organic fouling found in RO elements that treat wastewater is usually a complex mixture of biological and abiotic organic compounds. The methods commonly used for membrane fouling quantification are nonspecific and measure all organic compounds present as either TOC or TN. The analysis of the contribution of carbohydrates and proteins in the measured TOC and TN values can be linked to the proportion of EPS in the fouling. This allows a clearer understanding of whether the main source of the organic foulants is biologic or abiotic. Samples from the elements taken from a "trial with nutrients added" and a "trial with no nutrients added" were used to validate that the new method can distinguish between the foulants produced in a high biofouling environment (nutrients dosed) or a low biofouling environment (no nutrients).

When nutrients were added, the percentage of TOC and TN accounted for by the carbohydrate and protein was nearly 100%. However, the percentages were much lower when no nutrients were dosed, indicating that a greater fraction of the fouling was caused by abiotic organic compounds. Additionally, the concentration of bacteria measured by ATP was found to be much higher than when the feed water was dosed with nutrients.

The ability to determine the proportion of EPS in the TOC and TN results has shown to be useful to determine the source of the compounds present as membrane foulants. The method described in this chapter will enable industrial water treatment plants to easily quantify the proportion of biological fouling present versus the proportion of non-biological organic fouling. Once assessed, pretreatments, operating conditions, and cleaning protocols can be adjusted to tackle the primarily type of fouling occurring. Pretreatment optimization strategies, such as biocide dosage or nutrient limitation, could be implemented when biofouling is determined to be the main type of fouling. Additionally, chemical cleanings protocols can be adapted for the predominant type of foulant present, such as the use of sanitizers or protease-based enzymatic cleaners for biofouling. Pre-concentration protocols for water samples will be explored in the future, to adapt the method to characterize the feed water foulant composition. This will provide a method to monitor the removal of each particular foulant type after specific pretreatments steps.

References

[1]   F. Beyer, M. Rietman, A. Zwijnenburg, P. Van den Brink, J. S. Vrouwenvelder, M. Jarzembowska, J. Laurinonyte. Long-term performance and fouling analysis of full-scale direct nanofiltration (NF) installations treating anoxic groundwater. Journal of Membrane Science, 468 (2014) 339–348.

[2]   J. S. Vrouwenvelder, D. van der Kooij. Diagnosis, prediction and prevention of biofouling of NF and RO membranes. Desalination, 139 (2001) 65–71.

[3]   T. Nguyen, F. Roddick, L. Fan. Biofouling of water treatment membranes: A review of the underlying causes, monitoring techniques and control measures. Membranes, 2 (2001) 804–840.

[4]   M. Al-Ahmad, F. A. Aleem, A. Mutiri, A. Ubaisy. Biofouling in RO membrane systems part 1: Fundamentals and control. Desalination, 132 (2000) 173–179.

[5]   C. Dreszer, J. S. Vrouwenvelder, A. H. Paulitsch-Fuchs, A. Zwijnenburg, J. C. Kruithof, H. C. Flemming. Hydraulic resistance of biofilms. Journal of Membranes Science, 429 (2013) 436–447.

[6]   J. S. Vrouwenvelder, D. van der Kooij. Diagnosis of fouling problems of NF and RO membrane installations by a quick scan. Desalination, 153 (2000) 121–124.

[7]   A. Magic-Knezev, D. Van der Kooij. Optimisation and significance of ATP analysis for measuring active biomass in granular activated carbon filters used in water treatment. Water Research, 38 (2004) 3971–3979.

[8]   G. Naidu, S. Jeong, S. Vigneswaran, S. A. Rice. Microbial activity in biofilter used as a pretreatment for seawater desalination. Desalination, 309 (2013) 254–260.

[9]   K. Manish, S. Samer, W. R. Pearce. Investigation of seawater reverse osmosis fouling and its relationship to pretreatment type. Environmental Science Technology, 406 (2006) 2037–2044.

[10]   M. Filella. Understanding what we are measuring: Standards and quantification of natural organic matter. Water Research, 50 (2014) 287–293.

[11]   M. T. Khan, M. Busch, V. G. Molina, A. H. Emwas, C. Aubry, J. P. Croue. How different is the composition of the fouling layer of wastewater reuse and seawater desalination RO membranes? Water Research, 59 (2014) 271–282.

[12]   J. S. Vrouwenvelder, J. W. N. M. Kappelhof, S. G. J. Heijman, J. C. Schippers. Tools for fouling diagnosis of NF and RO membranes and assessment of the fouling potential of feed water. Desalination, 157 (2003) 361–365.

[13]   R. J. Barnes, R. R. Bandi, F. Chua, J. H. Low, T. Aung, N. Barraud, A. G. Fane. The roles of Pseudomonas aeruginosa extracellular polysaccharides in biofouling of reverse osmosis membranes and nitric oxide induced dispersal. Journal of Membranes Science, 466 (2014) 161–172.

[14]   C. Nuengjamnong, J. H. Kweon, J. Cho, C. Polprasert, K. H. Ahn. Membrane fouling caused by extracellular polymeric substances during microfiltration processes. Desalination, 179 (2005) 117–124.

[15]   S. S. Branda, F. Chu, D. B. Kearns, R. Losick, R. Kolter. A major protein component of the Bacillus subtilis biofilm matrix. Molecular Microbiology, 59 (2006) 1229–1238.

[16]   H. C. Flemming, J. Wingender. The biofilm matrix. Nature Reviews Microbiology, 8(9) (2010) 623–633.

[17]   Q. Wei, L. Ma. Biofilm matrix and its regulation in Pseudomonas aeruginosa. International Journal of Molecular Sciences, 14 (2013) 20983–21005.

[18]   B. W. Peterson, Y. He, Y. Ren, A. Zerdoum, M. R. Libera, P. K. Sharma, A. J. Van Winkelhoff. Viscoelasticity of biofilms and their recalcitrance to mechanical and chemical challenges. FEMS Microbiology Reviews, 39(2) (2015) 234–245.

[19]   J. S. Vrouwenvelder, S. A. Manolarakis, J. P. Van der Hoek, J. A. M. Van Paassen, J. M. C. Van Agtmaal, M. C. M. Van Loosdrecht. Quantitative biofouling diagnosis in full scale nanofiltration and reverse osmosis installations. Water Research, 42 (2008) 4856–4868.

[20]   J. A. Novitsky. Degradation of dead microbial biomass in a marine sediment. Applied and Environmental Microbiology, 52 (1986) 504–509.

[21]   M. L. Yallop, D. M. Paterson, P. Wellsbury. Interrelationships between rates of microbial production, exopolymer production, microbial biomass, and sediment stability in biofilms of intertidal sediments. Microbial Ecology, 39 (2000) 116–127.

[22]   W. A. M. Hijnen, E. R. Cornelissen, D. van der Kooij. Threshold concentrations of biomass and iron for pressure drop increase in spiral-wound membrane elements. Water Research, 45 (2011) 1607–1616.

[23]   M. Herzberg, S. Kang, M. Elimelech. Role of extracellular polymeric substances (EPS) in biofouling of reverse osmosis membranes. Environmental Science & Technology, 4312 (2009) 4393–4398.

[24]   L. O. Villacorte, M. D. Kennedy, G. L. Amy, J. C. Schippers. The fate of transparent exopolymer particles (TEP) in integrated membrane systems: Removal through pre-treatment processes and deposition on reverse osmosis membranes. Water Research, 43 (2009) 5039–5052.

[25] T. J. Stewart, J. Traber, A. Kroll, R. Behra, L. Sigg. Characterization of extracellular polymeric substances (EPS) from periphyton using liquid chromatography-organic carbon detection–organic nitrogen detection (LC-OCD-OND. Environmental Science and Pollution Research, 205 (2013) 3214–3223.

[26] B. G. Choi, J. Cho, K. G. Song. Correlation between effluent organic matter characteristics and membrane fouling in a membrane bioreactor using advanced organic matter characterization tools. Desalination, 309 (2013) 74–83.

[27] S. A. Huber, A. Balz, M. Abert, W. Pronk. Characterisation of aquatic humic and non-humic matter with size-exclusion chromatography–organic carbon detection–organic nitrogen detection (LC-OCD-OND. Water Research, 45 (2011) 879–885.

[28] M. T. Khan, C. L. De O. Manes, C. Aubry, L. Gutierrez. Kinetic study of seawater reverse osmosis membrane fouling. Environmental Science & Technology, 47 (2013) 10884–10894.

[29] S. E. Jensen, P. Hubrechts, B. M. Klein, K. R. Hasløv. Development and validation of an ATP method for rapid estimation of viable units in lyophilised BCG Danish 1331 vaccine. Biologicals, 36 (2008) 308–314.

[30] G. Massons-Gassol, G. Gilabert-Oriol, R. Garcia-Valls, V. Gomez, T. Arrowood. Development and application of an accelerated biofouling test in flat cell. Desalination and Water Treatment, 57 (2016) 23318–23325.

[31] A. C. Fonseca, R. S. Summers, A. R. Greenberg, M. T. Hernandez. Extracellular polysaccharides, soluble microbial products, and natural organic matter impact on nanofiltration membranes flux decline. Environmental Science & Technology, 41 (2007) 2491–2497.

[32] J. Cho, S. W. Hermanowicz, J. Hur. Effects of experimental conditions on extraction yield of extracellular polymeric substances by cation exchange resin. The Scientific World Journal, 2012 (2001) 1–6.

[33] A. Eldyasti, G. Nakhla, J. Zhu. Impact of calcium on biofilm morphology, structure, detachment and performance in denitrifying fluidized bed bioreactors (DFBBRs. Chemical Engineering Journal, 232 (2013) 183–195.

[34] S. R. Suwarno, X. Chen, T. H. Chong, V. L. Puspitasari, D. McDougald, Y. Cohen, A. G. Fane. The impact of flux and spacers on biofilm development on reverse osmosis membranes. Journal of Membrane Science, 405 (2001) 219–232.

[35] E. Zuriaga-Agustí, A. Bes-Piá, A. Mendoza-Roca. Influence of extraction methods on proteins and carbohydrates analysis from MBR activated sludge flocs in view of improving EPS determination. Separation and Purification Technology, 112 (2013) 1–10.

[36] M. Dubois, K. A. Gilles, J. K. Hamilton, F. Rebers. Colorimetric method for determination of sugars and related substances. Analytical Chemistry, 28 (1956) 350–356.

[37] T. Kawaguchi, A. Decho. A laboratory investigation of cyanobacterial extracellular polymeric secretions (EPS) in influencing $CaCO_3$ polymorphism. Journal of Crystal Growth, 240 (2000) 230–235.

[38] W. Hijnen, C. Castillo, A. Brouwer-Hanzens. Quantitative assessment of the efficacy of spiral-wound membrane cleaning procedures to remove biofilms. Water Research, 46 (2001) 6369–6381.

[39] X. Pan, J. Liu, D. Zhang. Comparison of five extraction methods for extracellular polymeric substances (EPS) from biofilm by using three dimensional excitation-emission matrix (3DEEM) fluorescence spectroscopy. Water SA, 36 (2010) 111–116.

[40] X. Zhang, P. Bishop. Biodegradability of biofilm extracellular polymeric substances. Chemosphere, 50 (2003) 63–69.

[41] P. Smith, R. I. Krohn, G. T. Hermanson, A. K. Mallia, F. H. Gartner, M. Provenzano, D. C. Klenk. Measurement of protein using Bicinchoninic acid. Analytical Biochemistry, 150 (1985) 76–85.

[42] C. Pellicer-Nàcher, C. Domingo-Félez, A. Mutlu. Critical assessment of extracellular polymeric substances extraction methods from mixed culture biomass. Water Research, 47 (2013) 5564–5574.

[43] A. Bujacz. Structures of bovine, equine and leporine serum albumin. Acta Crystallographica Section D: Biological Crystallography, 68(10) (2001) 1278–1289.

[44] W. Ying, N. Siebdrath, W. Uhl, V. Gitis, M. Herzberg. New insights on early stages of RO membranes fouling during tertiary wastewater desalination. Journal of Membrane Science, 466 (2014) 26–35.

[45] X. Zheng, R. Mehrez, M. Jekel, M. Ernst. Effect of slow sand filtration of treated wastewater as pretreatment to UF. Desalination, 249 (2009) 591–595.

Chapter 8
Biofouling starts to develop before pressure drop starts to increase

Correlation between pressure drop, adenosine triphosphate, and total organic fouling: assessing their impact on biofouling development in reverse osmosis membranes

This research shows how, despite biofilm growth initially not being noticed by an increase of pressure drop, it might have already started forming. This is confirmed by the measurement of adenosine triphosphate (ATP) accumulation on the membrane, where it is observed that ATP concentration, which is found in bacteria, starts increasing very fast at the beginning of the trial, when pressure drop stills remains quite stable. This research also shows how total organic carbon (TOC) starts accumulating quickly on the membrane at the beginning of the trial. This indicates the beginning of the organic fouling period. This TOC accumulation on the membrane matches well with the mechanism by which biofilms are formed, as for a biofilm to start growing, it needs a substrate. This is referred to as the conditioning layer. This research also shows how ATP measurement on a membrane might be used as an early indicator of biofouling. This thus helps to identify when the biofilm starts to form before the early biofilm formation leads to a biofilm that is big enough to start causing an increase in pressure drop, thereby leading to the problem of biofouling in an RO system. It is worth noting that the ATP starts to accumulate after the TOC has already accumulated.

8.1 Introduction

Water scarcity is recognized as one of the main threats that mankind is facing globally [1]. RO membrane technology has developed as a promising, cost-effective technology to remove contaminants from non-potable waters and provide fresh water supply to meet the growing demand [2]. RO elements, however, can suffer from progressive loss of performance, when treating challenging waters, due to fouling [3]. Of all the fouling types, biofouling is the most complex to manage in RO water treatment systems [4]. It

Acknowledgments: The author would like to acknowledge the contributors of this chapter: Guillem Gilabert-Oriol, Veronica Gomez, Gerard Massons-Gassol, and Tina Arrowood. The author would also like to thank all DuPont Water Solutions teams for their support in doing this research, and specifically Javier Dewisme and Javier de la Fuente for their reliable work in operating the pilot plant in the Vilaseca WWTP and Mireia Font and Patricia Carmona for the excellent analytical support.

This chapter is in the process of being submitted for publication, and was presented at the Aachener Membran Kolloqium (AMK) conference held in Aachen (Germany) in November 2014.

https://doi.org/10.1515/9783111639079-008

occurs when bacteria colonize and form biofilms in the feed channel of the RO elements, causing increased friction for water flow. This causes the differential pressure (dP) to increase [5], leading to hydraulic imbalance and, if not controlled, can damage the element. Additionally, biofilms can affect membrane transport properties, as the polymeric film formed on the membrane surface decreases the overall water permeability [6]. Each of these effects increase the energy of operation and leads to frequent system shutdowns for chemical cleanings to regain membrane performance. The high pH conditions needed to remove biofilms during cleaning can result in membrane hydrolysis and shorten the useful life of the element. Therefore, in total, system productivity, chemical usage, membrane life, and energy, each contribute to higher cost of water production when biofouling occurs. Mechanisms to control biofouling are needed to enable long-term performance when treating water with high contamination levels [7].

The objective of this experiment is to understand the correlation between the pressure drop (dP) of an RO element, which is a universal measure to monitor biofouling growth in an RO membrane, with the ATP and TOC present on a membrane, once it is autopsied.

8.2 Materials and methods

8.2.1 Wastewater characterization

Vila-seca wastewater treatment plant (WWTP) treats municipal wastewater from Vila-seca, La Pineda, and Salou villages. The intake of the water is from the effluent of the clarifier, which is then pretreated using a membrane bioreactor (MBR). The average ionic and organic characteristics of the ultrafiltered water used for the RO experiments are summarized in Table 8.1.

Table 8.1: Wastewater feed characterization.

Parameter	Average concentration
ATP	0.03 ± 0.02 ng/L
TOC	8 ± 1.5 mg/L
Calcium	135 ± 16 mg/L
Chloride	509 ± 48 mg/L
Nitrate	63 ± 26 mg/L
Sodium	308 ± 30 mg/L
Sulfate	290 ± 49 mg/L
Conductivity	$2{,}408 \pm 442$ µS/cm
pH	7.3 ± 0.2

8.2.2 Pilot plant

Different trials were performed at the Fouling Resistant Elements Pilot Plant (FREPP), located in the Vila-seca wastewater treatment plant (WWTP). A single high-pressure centrifugal pump is used to feed the eight RO membranes, as shown in Figure 8.1. To match the flux and recovery of elements having different permeability, each pressure vessel has a needle valve in the permeate port to apply individual backpressure to each element. In the first trials, backpressure was registered manually using analog pressure indicators. On a later phase, automatic pressure transmitters were installed with the intention to track the changes on backpressure on a continuous mode. The plant can also operate on recirculation mode using activated carbon filters, which are used during the stabilization phase in order to capture any organics that the elements might leach to the feed water. This experiment was performed at the Global Water Technology Center that DuPont Water Solutions has in Tarragona, Catalonia, Spain.

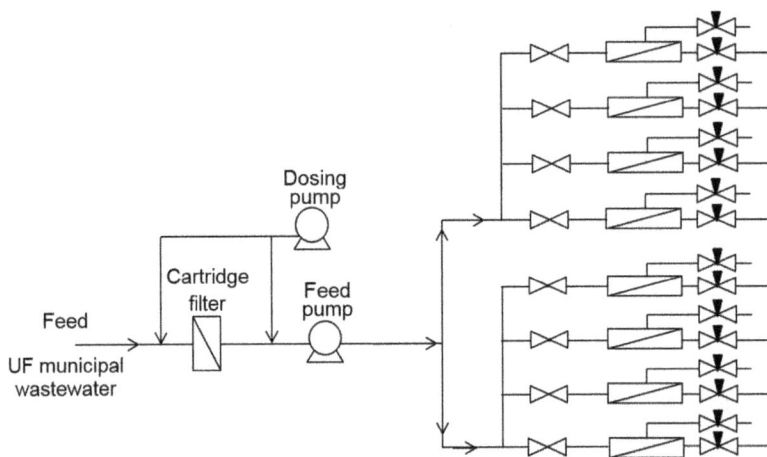

Figure 8.1: Fouling Resistant Elements Pilot Plant scheme.

8.2.3 Experimental setup

Before starting the test, the whole unit was sanitized to make sure no fouling from any previous trial was present, following the steps described in Table 8.2, using 2,2-dibromo-3-nitrilopropionamide (DBNPA), as described in the sanitization protocol. Cartridge filters were used as is usual in any water treatment plant. Sodium metabisulfite (SMBS) was not used. In order to accelerate the biofouling development, 0.1 mg/L of sodium acetate was dosed as an organically assimilable nutrient. Sodium acetate solution was buffered at pH 12 to make sure the stock solution being dosed could be properly preserved from any unintended microbial growth.

Seven single DuPont™ FilmTec™ BW30 2514 membrane elements were installed in parallel, with the aim of gradually periodically removing a membrane during the operation. Once a membrane is removed, it is autopsied and the ATP and TOC present on each membrane surface are analyzed, and their values were matched against the membrane pressure drop values.

Table 8.2: Sanitization protocol.

Step	Task
1	Flushing with deionized water to eliminate visible bacteria sludge
2	1–2 h circulation with NaOH of pH 12–13 at temperature of 25–35 °C
3	Flushing with deionized water
4	1 h recirculation with 60 ppm DBNPA
5	Flushing with deionized water

8.3 Results and discussion

8.3.1 Pressure drop

The pressure drop evolution over time is shown in Figure 8.2. Different membranes where autopsied and removed at different times, so that the ATP and TOC values could be matched with the differential pressure that each membrane showed.

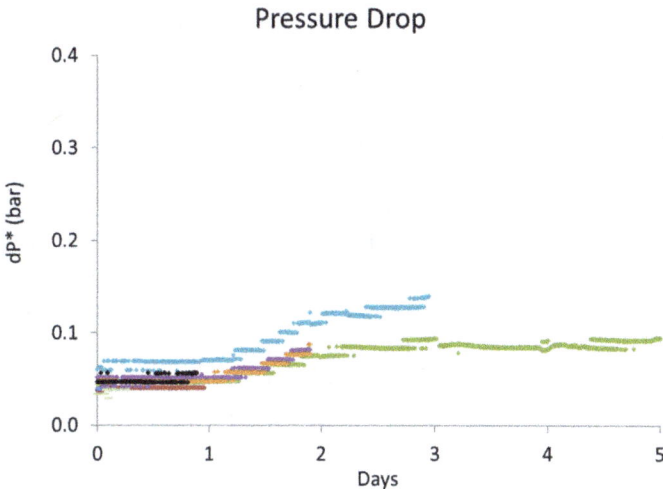

Figure 8.2: Pressure drop evolution.

The pressure drop corresponding to each membrane that was removed can be seen in Figure 8.3. In this plot, the typical profile of biofouling can be observed, where initially the pressure drop is stable and suddenly, it starts to increase exponentially. In this experiment, the pressure drop started to increase after the first day, as nutrients were dosed to accelerate biofouling.

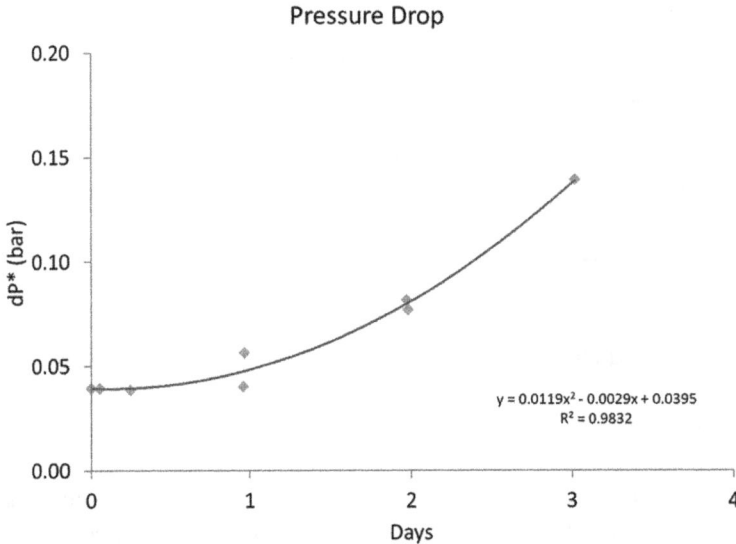

Pressure Drop

$y = 0.0119x^2 - 0.0029x + 0.0395$
$R^2 = 0.9832$

Figure 8.3: Pressure drop of each membrane analyzed.

8.3.2 ATP

The ATP evolution over time is shown in Figure 8.4. These values correspond to each membrane autopsied from the system. It is worth noticing that while the pressure drop increase over time tends to follow an exponential increase pattern, for ATP, the plot resembles a logarithmic pattern. This means that initially, there is a rapid growth of ATP, and then this increase flattens out. If the pressure drop plot over time is compared with the ATP evolution over time, it is very interesting to realize that while initially the pressure drop hardly increased, ATP experiences a large increase for the same initial period. This is attributed to the fact that before establishing a biofilm, the bacteria need to attach to the membrane, and as the ATP represent the bacteria, increase of ATP is observed while the pressure drops remains quite stable. After the bacteria have built a sufficiently big biofilm, ATP does not increase that much, as the biofilm is mainly composed of extracellular polymeric substances.

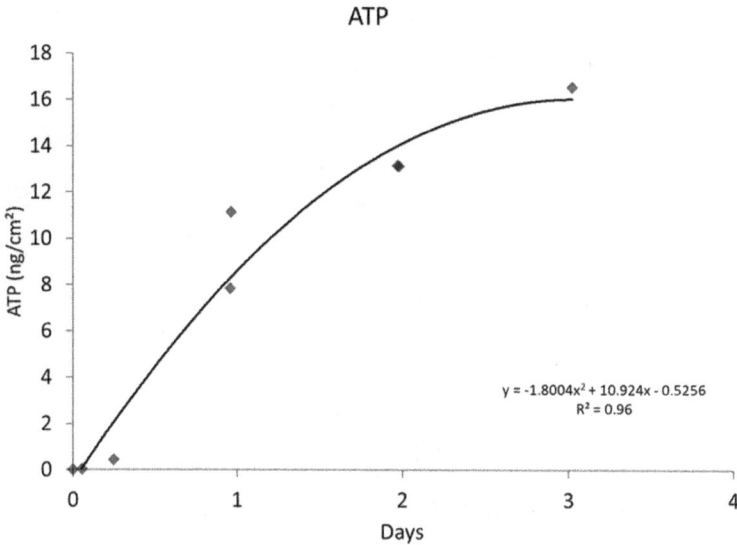

Figure 8.4: ATP of each membrane analyzed.

The relationship between the pressure drop and ATP can be more clearly observed in Figure 8.5. In this plot, while the pressure drop (dP) hardly increases and stays around 0.04 bar, ATP experiences a big increase, going from 0 to almost 8 ng/cm^2.

8.3.3 TOC

The TOC evolution over time is shown in Figure 8.6. This plot is interesting because it shows how TOC quickly accumulates on the membrane during the start of the operation, and then it tends to plateau. This increase resembles, like ATP, a logarithmic trend. This is consistent with an organic fouling accumulation during the first day of operation, when biofouling has not yet been noticed, as pressure drop remains flat over time. This initial build up of TOC over time is what leads to organic fouling, and also, it conditions the membrane with organics, which are likely used by bacteria to grow and start developing a biofilm. It is also noted that ATP starts to increase after TOC has already accumulated.

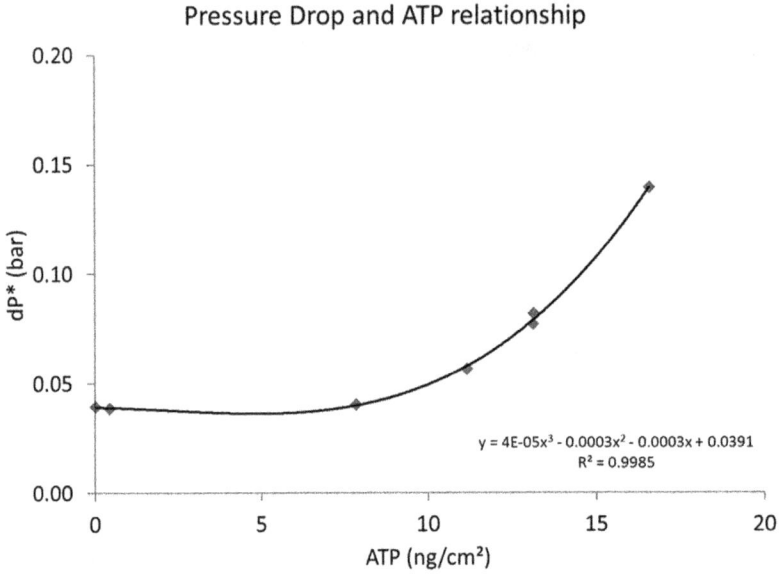

Figure 8.5: Relationship between pressure drop and ATP.

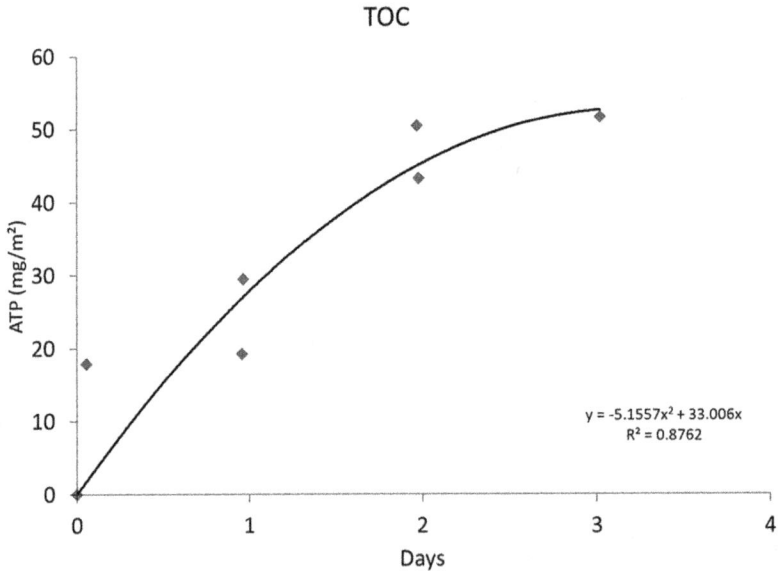

Figure 8.6: TOC of each membrane analyzed.

8.4 Conclusions

This research shows that, although a biofilm might not be noticed by an increase in pressure drop, it might be already started forming. This is confirmed by the measurement of ATP accumulation on the membrane, where it is observed that ATP, which is found in bacteria, starts increasing very fast at the beginning of the trial, when pressure drop stills remains quite stable. This research also shows how TOC starts accumulating quickly on the membrane at the beginning of the trial, corresponding to an initial organic fouling period. This TOC accumulation on the membrane matches well with the mechanism by which biofilms are formed, as for a biofilm to start growing, it needs a substrate. This is referred to as the conditioning layer. This research also shows how ATP measurement on a membrane might be used as an early indicator of biofouling. It might help to identify when the biofilm is starting to form before this early biofilm formation leads to a biofilm that is big enough to start causing an increase in pressure drop, thereby leading to the problem of biofouling in an RO system. It is worth noting that the ATP starts to increase after the TOC has already accumulated.

References

[1] C. Fritzmann, J. Löwenberg, T. Wintgens, T. Melin. State-of-the-art of reverse osmosis desalination. Desalination, 216 (2007) 1–76.
[2] L. F. Greenlee, D. F. Lawler, B. D. Freeman, B. Marrot, P. Moulin. Reverse osmosis desalination: Water sources, technology, and today's challenges. Water Research, 43 (2009) 2317–2348.
[3] K. P. Lee, T. C. Arnot, D. Mattia. A review of reverse osmosis membrane materials for desalination – development to date and future potential. Journal of Membrane Science, 370 (2011) 1–22.
[4] H. C. Flemming. Biofouling in water systems–cases, causes and countermeasures. Applied Microbiology and Biotechnology, 59 (2002) 629–640.
[5] C. Sun, L. Fiksdal, A. Hanssen-Bauer, M. B. T. Rye, T. Leiknes. Characterization of membrane biofouling at different operating conditions (flux) in drinking water treatment using confocal laser scanning microscopy (CLSM) and image analysis. Journal of Membrane Science, 382 (2011) 194–201.
[6] S. R. Suwarno, X. Chen, T. H. Chong, V. L. Puspitasari, D. McDougald, Y. Cohen, S. A. Rice, A. G. Fane. The impact of flux and spacers on biofilm development on reverse osmosis membranes. Journal of Membrane Science, 405 (2012) 219–232.
[7] A. Matina, Z. Khana, S. M. J. Zaidia, M. C. Boyceb. Biofouling in reverse osmosis membranes for seawater desalination: Phenomena and prevention. Desalination, 281 (2011) 1–16.

Chapter 9
Biofouling mainly happens in the lead elements

Investigating biofouling distribution in reverse osmosis membrane systems: Insights into cleaning efficiency and system performance

This research shows how a smart distribution of different types of feed spacers in a pressure vessel can mitigate biofouling, with a focus on the biofouling that mainly develops in the lead elements. Results highlight the difficulty of cleaning back to the initial pressure drop once biofouling develops, as it is very difficult to fully remove a biofilm inside a reverse osmosis membrane. Caustic cleaning provides the highest cleaning effectiveness among the different strategies. It also shows that biocide cleaning can reduce the number of bacteria in a biofilm, but does not prevent biofouling. It is also observed that biofouling can be easily brushed or removed by stirring in a beaker from a membrane and feed spacer. Finally, this trial shows the two classical fouling phases, the first organic fouling phase and the later biofouling phase, which grows over the already organically fouled membrane.

9.1 Introduction

One of the main hurdles for membrane filtration using challenging feed water is biofouling [1–2]. Biofouling is the growth of biofilm within the module that causes flux decline, an increase in energy consumption, and chemical cleaning frequency, thereby lowering the membrane lifetime [3]. Biofilm is defined as a complex assemblage of microbial communities attached to a substratum and to each other, by means of a dense, self-produced biopolymer matrix of extracellular polymeric substances (EPSs). This hydrogel is mainly composed of polysaccharides and proteins and has high water content [4]. EPS has been recognized as the main contributor to the operational performance decline of elements suffering from biofouling [5]. A comparable effect is caused by the transparent exopolymer particles (TEPs), which share similar characteristics with EPS [6]. TEP has recently raised awareness, as it is commonly found during algal blooms and causes a severe decrease in membrane permeability [7].

Acknowledgments: The author would like to acknowledge the contributors of this chapter: Guillem Gilabert-Oriol, Gerard Massons, Diana Dubert, Christopher Rieth, Carolina Martínez, Patricia Carmona, Veronica Gómez, Javier Dewisme, Javier de la Fuente, Steve Jons, Jon Johnson, and Tina Arrowood. The author would also like to thank all members of the DuPont Water Solutions team for their support on doing this research.

This chapter is in the process of being submitted for publication, and was presented at the American Dairy Science Association (ADSA) conference held in Pittsburg (United States of America) in June 2017.

https://doi.org/10.1515/9783111639079-009

The objective of this trial is to assess the performance of two different membrane configurations that comprise membranes with different feed spacer thicknesses across the pressure vessel. The idea is that the wider feed spacer that has the lowest pressure drop can be placed in the lead elements, while the thinner feed spacer can be placed in the tail part of the pressure vessel, thus helping with the mixing and helping to mitigate inorganic fouling.

The typical trade-off that is taken when following this approach is that a wider feed spacer usually leads to a smaller membrane active area, while a smaller feed spacer usually leads to a higher active area. The idea is that the configuration of thicker spacers in the front and thinner spacers in the tail positions of the pressure vessel would yield the desired total active area for the system.

This experiment will also attempt to investigate the fouling distribution in terms of biological and organic fouling in the different positions of the pressure vessel.

Finally, an attempt will be made to assess how easy or difficult it is to clean the remaining biofouling that the membrane has after the last cleaning in place (CIP) is performed.

9.2 Materials and methods

9.2.1 Wastewater characterization

Vila-seca wastewater treatment plant treats municipal wastewater from Vila-seca, La Pineda, and Salou villages. The intake of the water is from the effluent of the secondary clarifier, which is then pretreated using two DuPont™ Ultrafiltration SFP-2880 XP modules. The average ionic and organic characteristics of the ultrafiltered water used for the reverse osmosis experiments are summarized in Table 9.1.

Table 9.1: Wastewater feed characterization.

Parameter	Average concentration
ATP	0.03 ± 0.02 ng/L
TOC	8 ± 1.5 mg/L
Calcium	135 ± 16 mg/L
Chloride	509 ± 48 mg/L
Nitrate	63 ± 26 mg/L
Sodium	308 ± 30 mg/L
Sulfate	290 ± 49 mg/L
Conductivity	$2{,}408 \pm 442$ µS/cm
pH	7.3 ± 0.2

9.2.2 Wastewater pilot plant and experimental setup

The Water Reuse Pilot Plant (WRUPP) is located at the Global Water Technology Center (GWTC) that DuPont has in Tarragona, Catalonia, Spain. This plant consists of two parallel reverse osmosis lines, each with its own cartridge filter and high-pressure pump. Each line can have up to two stages. The first stage consists of a 4-inch spiral-wound membrane stage, while the second stage, because of hydraulic considerations, consists of a 2.5-inch spiral-wound membrane stage. A diagram of the pilot plant and the experimental setup, together with the different combinations of the feed spacers, can be found in Figure 9.1. So, the first line has the novel staged configuration, with the first two membranes having 46-mil spacers, the third to sixth membranes having the 34-mil LDP spacers, and the second stage, the 28-mil spacer. The second line has the standard configuration with 34-mil LDP spacers. It should be noted that all the feed spacers are referred in mils (0.001 inches), and the abbreviation LDP refers to low pressure drop.

Spiral-wound DuPont™ FilmTec™ BW30HRLE reverse osmosis membranes are used for this experiment. Both lines have the same total active area. The system operates at a constant average flux of 17 L/m²h and recovery of 62%. A CIP is performed when the feed-side pressure drop reaches 3.5 bar.

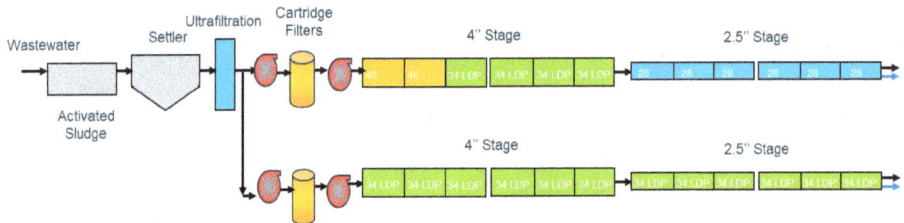

Figure 9.1: Wastewater pilot plant diagram.

9.2.3 Membrane fouling simulator skid

The membrane fouling simulator (MFS) skid is a pilot plant consisting of up to six MFSs in parallel. This pilot plant allows for quickly studying biofouling. A schematic of this plant can be seen in Figure 9.2. This plant is used to analyze the cleanability potential of the membrane coupons extracted from the end of the experiment that is performed with the wastewater pilot. It should be noted that the experiment ended with a cleaning-in place, so these membranes were considered cleaned.

To perform this cleaning study, membrane samples have been extracted from the reverse osmosis membrane elements. The membrane and the feed spacer coupon are cut into a 20 cm × 5 cm rectangle. Then, the coupon is placed in the skid, where the cross-flow velocity is gradually increased above the wastewater pilot operating velocity of 0.11 m/s and the CIP flushing velocity of 0.21 m/s. After each flushing, the system flow is returned to 16 L/h (0.13 m/s) to measure the pressure drop and assess the cleaning efficiency.

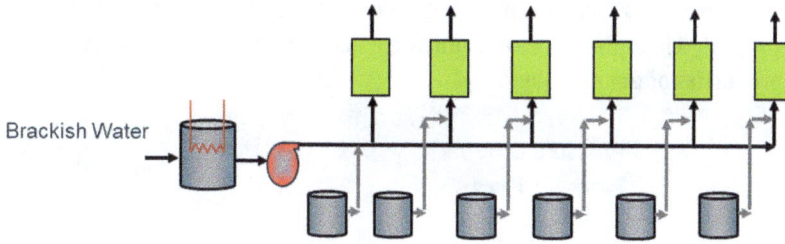

Figure 9.2: Membrane fouling simulator skid.

9.2.4 Cleaning protocol in the wastewater plant

When the system pressure drop reaches 3.5 bar, the cleaning procedure listed below is followed according to Table 9.2. It should be noticed that, first, a biocide cleaning step is performed using 2,2-dibromo-3-nitrilopropionamide (DBNPA). Then, a caustic cleaning step is carried out. Finally, an acid cleaning step is performed.

Table 9.2: Cleaning protocol.

Step	Cleaning	Details
1	Biocide cleaning step	– Performing 2 h recirculation of 100 ppm DBNPA pH 4 (flushing with RO permeate) – Overnight soaking of 100 ppm DBNPA pH 4 (flushing with RO permeate)
2	Caustic cleaning step	– Performing 2 h recirculation with pH 13 at 35 °C (flushing with RO permeate) – Overnight soaking with pH 13 (flushing with RO permeate)
3	Acid cleaning step	– Performing 2 h recirculation with pH 2 (flushing with RO permeate) – Overnight soaking with pH 2 (flushing with RO permeate)

9.3 Results and discussion

9.3.1 Biofouling prevention study

9.3.1.1 Pressure drop

The evolution of pressure drop over time, indicating the amount of biofouling that the membranes are developing, is shown in Figure 9.3. It is interesting to realize that the cleaning protocol that is followed cannot recover the pressure drop to its initial value of 0.5 bar. The minimum pressure drop value that it can achieve is 1.5 bar. This is referred as the non-cleanable biofilm. It is also noticeable that this novel feed spacer arrangement shows a 20% lower pressure drop initially and an initial benefit of reducing CIP frequency by 30% from the onset time. It should also be noticed that the inability to properly clean the reverse osmosis elements might be detrimental to the antifouling properties of the module.

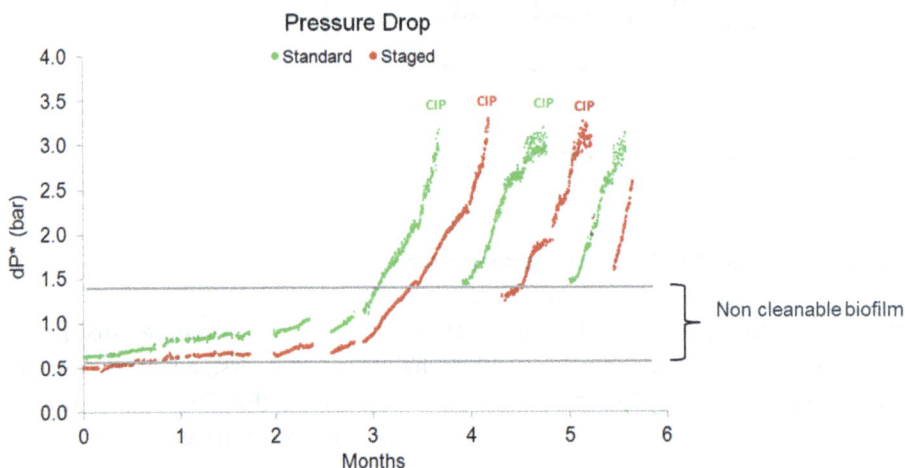

Figure 9.3: Pressure drop evolution.

9.3.1.2 Water permeability

The water permeability evolution over time, represented by the A-value, can be seen in Figure 9.4. In this diagram, it can be seen that the initial permeability drop caused the first organic fouling period. It can also be noticed when the biofouling starts to develop – between the second and third month. This corresponds to the biofouling period.

9.3.2 Where biofouling develop

9.3.2.1 Pressure drop

This wastewater experimental plant has the peculiarity that the pressure drop can be measured between the lead elements, which correspond to positions 1–3, and the tail elements, which correspond to positions 4–6. This pressure drop distribution between the lead and tail elements can be observed in Figure 9.5. From this diagram, it can be seen that biofouling mainly develops in lead elements, while hardly any biofouling happens in the tail elements. This points to the fact that once a biofilm is established

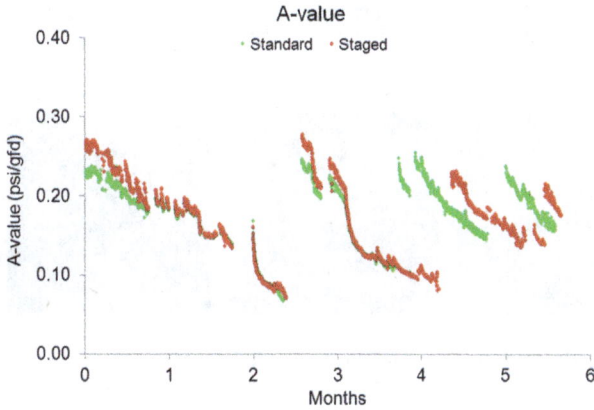

A-value

Standard · Staged

Figure 9.4: Water permeability (*A*-value) evolution.

in the lead elements, the bacteria will consume the most limiting organic assimilable nutrients, and therefore, biofouling will be attenuated in the downstream elements.

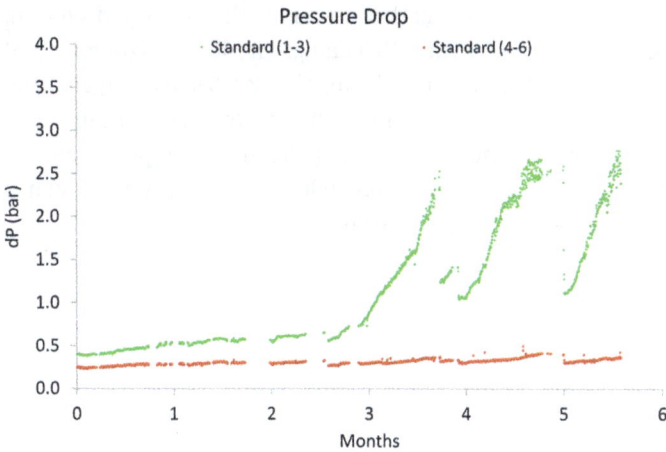

Pressure Drop

· Standard (1-3)   · Standard (4-6)

Figure 9.5: Pressure drop evolution in the tail and lead elements.

9.3.2.2 Fouling visual observation

After performing the last chemical cleaning in both lines, elements from different positions were autopsied. Figure 9.6 shows the membrane and feed spacer of the first, second, fourth, and twelfth elements. When looking at the fouling distribution across the pressure vessel, it is interesting to realize that the biofilm is only visually appreciated in the first element's feed spacer.

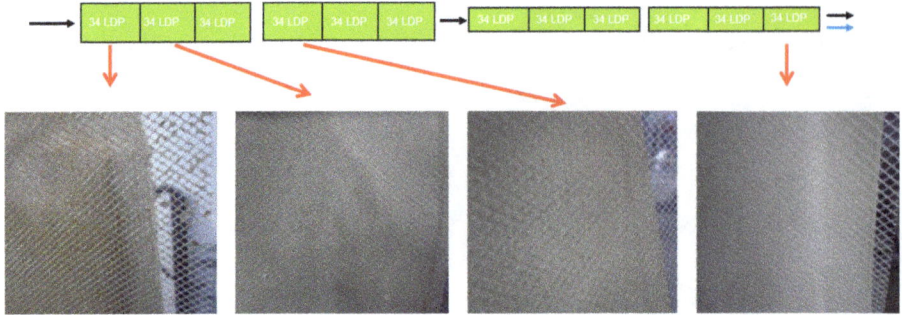

Figure 9.6: Membrane and feed spacer of the first, second, fourth, and twelfth elements.

9.3.2.3 Fouling distribution

To perform a proper quantification of the fouling distribution across the pressure vessel (PV) in each one of these autopsied elements, the organic and inorganic fractions present in the fouling of the first, second, and fourth elements are quantified in Figure 9.7. These autopsies are also performed after a chemical cleaning (CIP) to better understand why there seems to be a non-cleanable biofilm that prevented the pressure drop from going back to its initial value after a chemical cleaning. The loss-on-ignition (LOI) method is used to perform this quantification in the distribution between organic and inorganic fouling. From this plot, it can be concluded that biofouling is accumulated mainly on lead elements. It should also be noted that the autopsies performed on the second stage elements could not be analyzed due to low fouling. This conclusion is in agreement with what is assumed in the scientific literature [8].

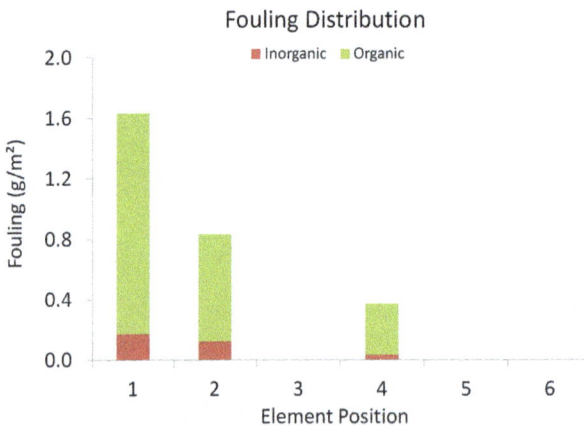

Figure 9.7: Fouling distribution across the pressure vessel.

9.3.2.4 Total organic carbon

The distribution of the total organic carbon (TOC) can be found in Figure 9.8. From this plot, high TOC levels confirm that the biofilm is mainly fouling the lead elements.

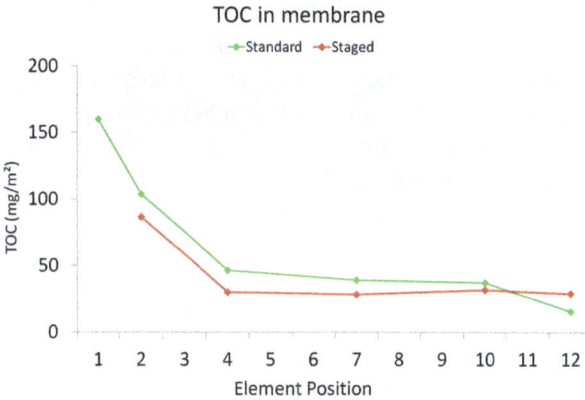

Figure 9.8: Total organic carbon distribution.

9.3.2.5 Total nitrogen

The distribution of the total nitrogen (TN) can be found in Figure 9.9. From this plot, it can also be seen that the high TN levels confirm that the biofilm is mainly fouling the lead elements.

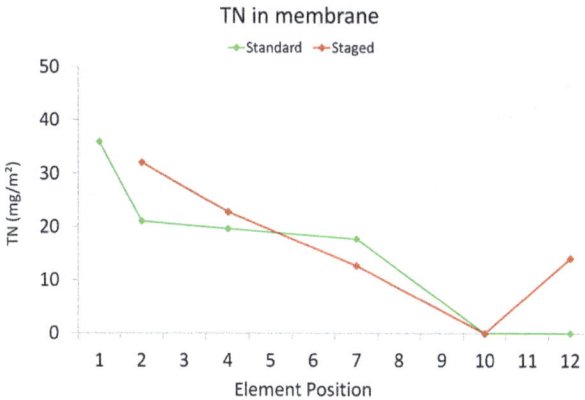

Figure 9.9: Total nitrogen distribution.

9.3.2.6 ATP analysis

Finally, the distribution of adenosine triphosphate (ATP) is also analyzed in Figure 9.10. In this plot, it can be observed that, most likely because of adding

2,2-dibromo-3-nitrilopropionamide (DBNPA) during the chemical cleaning, very low ATP values were found. Biocide significantly reduces bacteria, represented by ATP, but the remaining residue is almost sterilized biofilm, mainly composed of EPSs. It should be noted that a biofilm will never be fully devoid of bacteria, as some bacteria can shelter inside the EPS matrix. This is especially important as even if a single bacterium survives, if it receives enough food, it will be enough for this bacterium to replicate and start growing the biofilm again. As a conclusion, it can be summarized that biocide cleaning can reduce the number of bacteria in a biofilm, but does not prevent biofouling. It should be noted that most ATP analyses are under the limit of quantification (LOQ) of 0.015 ng/cm^2.

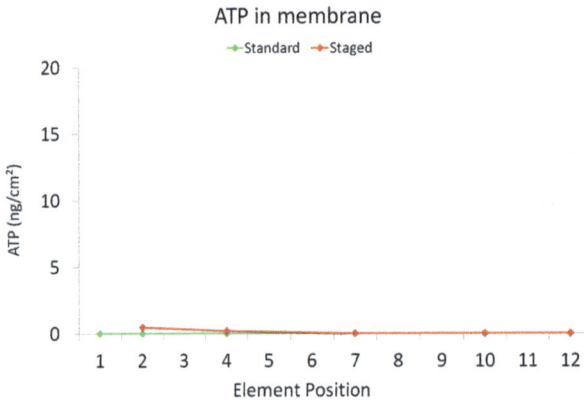

Figure 9.10: ATP distribution.

9.3.2.7 Visualization of biofouling on the membrane

The Fortilife™ Director™ Biofilm Visualization Tool method was used to visualize and quantify the portion of a membrane covered by a biofilm, as can be seen in Figure 9.11. These images again confirm that biofouling is mainly accumulated in the lead elements above the membrane.

9.3.2.8 Ease of scraping biofilm

To investigate why a fraction of biofouling is not eliminated after a chemical cleaning, manual scraping with a cotton swab was performed. The results of this scraping can be observed in Figure 9.12. From this small experiment, it can be observed that biofilm is easy to detach from the membrane when it is scraped. An experiment performed later and shown in this chapter, where the feed spacer that has biofilm is stirred in a beaker, also shows how easily it detaches from the feed spacer. This might indicate that in fact, biofilm can be easy to remove, but inside a reverse osmosis membrane, the feed spacer, together with the membrane surface, create low velocity

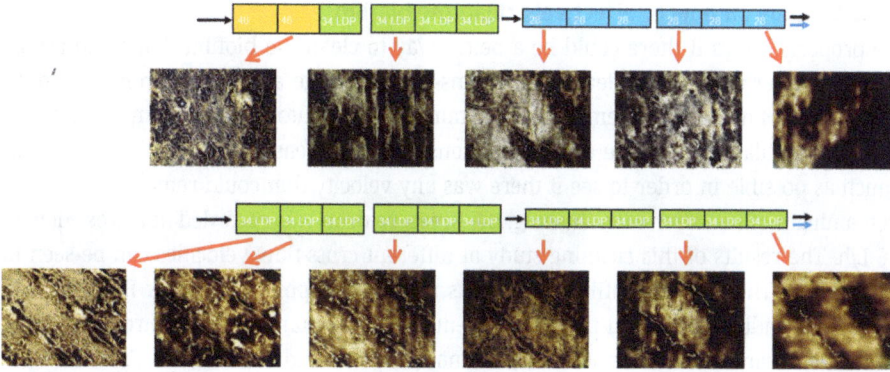

Figure 9.11: Visual fouling distribution.

zones, where the biofilm is sheltered and cannot be properly flushed away with water or cleaning chemicals. The narrow path inside the feed-concentrate channel might also play a role in making it more difficult for the biofilm to be flushed away. Also, the fact that the feed spacer is fixed inside the membrane and cannot be removed, also makes it more difficult for the biofilm to be removed.

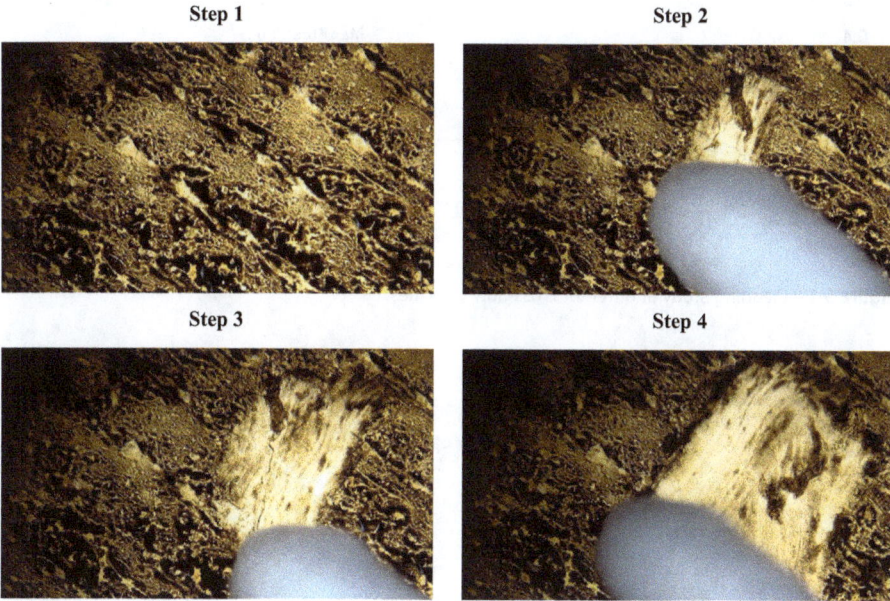

Figure 9.12: Scraping the biofilm.

9.3.2.9 Study on how to effectively clean the biofilm

To properly assess if there could be a better way to clean the biofilm that could not be removed after a chemical cleaning, coupons of membrane and spacer from a used reverse osmosis membrane element were obtained during autopsy and put in operation in the MFS simulation skid. The experiment consisted of increasing the flushing velocity as much as possible in order to see if there was any velocity that could remove most of the remaining biofilm. After each flushing, the pressure drop was recorded at a cross-flow of 16 L/h. The results of this cleaning study at different cross-flow velocities can be seen in Figure 9.13. The "Initial Biofilm" tag marks the starting point, the "Max Flushing" tag marks the ending point, and the "Clean Membrane" tag marks the pressure drop of the same membrane and feed spacer that are manually brushed and cleaned. From this diagram, it can be seen that despite an increase of four times in the cross-flow velocity, only an additional 40% of the non-cleanable biofilm can be cleaned. The visual observation of the membrane and feed spacer after this maximum flushing test is shown in Figure 9.14.

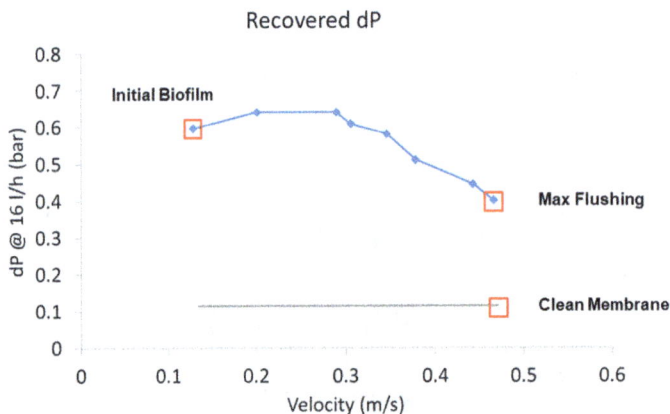

Figure 9.13: Cleaning study at different cross-flow velocities.

Figure 9.14: Membrane and feed spacer before and after the max flushing test, and manually brushed.

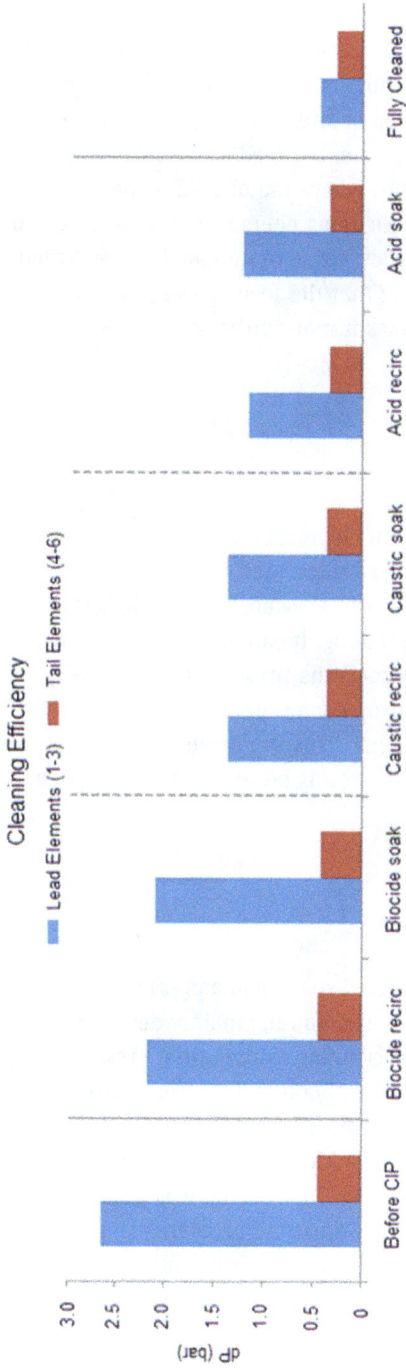

Figure 9.15: Cleaning study in the wastewater pilot.

9.3.2.10 Beaker cleaning study

Finally, an experiment performed where the feed spacer that has biofilm is manually and gently stirred in a beaker. This also shows how easily it detaches from the feed spacer. This might indicate, as has been explained before, that in fact, biofilm is easy to remove. But, inside a reverse osmosis membrane, the feed spacer, together with the membrane surface, creates low velocity zones, where the biofilm is sheltered and cannot be properly flushed away with water or cleaning chemicals. The narrow path inside the feed-concentrate channel might also play a role in making it more difficult for the biofilm to be flushed away. Also, the fact that the feed spacer is fixed inside the membrane and cannot be removed also makes it more difficult for the biofilm to be removed.

9.3.3 Cleaning study

In order to properly understand the effect of each chemical cleaning step performed in the wastewater pilot, pressure drop is measured after each cleaning step. The results are shown in Figure 9.15. From this study, it can be seen that caustic cleaning is the most effective cleaning step in reducing biofouling. It can also be concluded that overnight soaking and acid cleanings hardly impacts the pressure drop recovery. Finally, as has already been seen in the pressure drop evolution data, this cleaning protocol cannot recover the pressure drop below 1.5 bar. This gap between the maximum recovered pressure drop and the initial pressure drop is herein referred as the irreversible biofouling or non-cleanable biofilm.

9.4 Conclusions

This research shows how a smart distribution of feed spacers in a pressure vessel can reduce the impact of biofouling. It also shows that biofouling mainly develops in the lead elements. It also highlights the difficulty of cleaning to the initial pressure drop once biofouling develops. It shows that it is very difficult to fully remove a biofilm inside a reverse osmosis membrane. It also points out that caustic cleaning provides the highest cleaning effectiveness. It also shows that biocide cleaning can reduce the number of bacteria in a biofilm but does not prevent biofouling. It is also observed that biofouling can be easily brushed or removed by stirring in a beaker from a membrane and feed spacer. Finally, this trial shows the two classical fouling phases, the first organic fouling phase, and the later biofouling phase that grows over the already organically fouled membrane.

References

[1]   Y. N. Wang, C. Y. Tang. Fouling of nanofiltration, reverse osmosis, and ultrafiltration membranes by protein mixtures: The role of inter-foulant-species interaction. Environmental Science & Technology, 45(15) (2011) 6373–6379.

[2]   S. R. Suwarno, X. Chen, T. H. Chong, V. L. Puspitasari, D. McDougald, Y. Cohen, S. A. Rice, A. G. Fane. The impact of flux and spacers on biofilm development on reverse osmosis membranes. Journal of Membrane Science, 405 (2012) 219–232.

[3]   M. Jafari, A. D'haese, J. Zlopasa, E. R. Cornelissen, J. S. Vrouwenvelder, K. Verbeken, C. Picioreanu. A comparison between chemical cleaning efficiency in lab-scale and full-scale reverse osmosis membranes: Role of extracellular polymeric substances (EPS. Journal of Membrane Science, 609 (2020) 118189.

[4]   C. Dreszer, J. S. Vrouwenvelder, A. H. Paulitsch-Fuchs, A. Zwijnenburg, J. C. Kruithof, H.-C. Flemming. Hydraulic resistance of biofilms. Journal of Membrane Science, 429 (2013) 436–447.

[5]   M. Jafari, A. D'haese, J. Zlopasa, E. R. Cornelissen, J. S. Vrouwenvelder, K. Verbeken, C. Picioreanu. A comparison between chemical cleaning efficiency in lab-scale and full-scale reverse osmosis membranes: Role of extracellular polymeric substances (EPS. Journal of Membrane Science, 609 (2020) 118189.

[6]   N. Saeidi, M. Rzechowicz, M. F. Siddiqui, S. W. Teng, A. G. Fane, H. Winters. The potential role of transparent exopolymer particles (TEP) and their pseudo-precursors on biofouling of RO water treatment membranes. Desalination and Water Treatment, 182 1–12.

[7]   S. Meng, R. Wang, K. Zhang, X. Meng, W. Xue, H. Liu, D. Liang, Q. Zhao, Y. Liu. Transparent exopolymer particles (TEPs)-associated protobiofilm: A neglected contributor to biofouling during membrane filtration. Frontiers of Environmental Science & Engineering, 15(4) (2021) 1–10.

[8]   E. M. Hoek, T. M. Weigand, A. Edalat. Reverse osmosis membrane biofouling: Causes, consequences and countermeasures. Npj Clean Water, 5(1) (2022) 45.

Chapter 10
Biocides do not fully prevent biofouling

This chapter aims to assess the impact of shock-dosing 2,2-dibromo-3-nitrilopropionamide (DBNPA), a nonoxidizing biocide in spiral-wound reverse osmosis membranes. In this study, it can be seen that while dosing the biocide initially helped in delaying and mitigating the biofouling development, after a certain time, biofouling still developed naturally. So, after 5 months of operation, biofouling starts to develop normally. The explanation for this might be that biocide has certainly an effect in eliminating bacteria, but once the biofilm is big enough, bacteria can shelter inside the EPS, and therefore biocide might stop being effective. This study also showed how a fouling-resistant membrane shows higher resilience over time, in terms of losing less water permeability over time. This also highlights the importance of having membrane chemistries that are fouling resistant. Dosing biocide is a useful method to study the first organic fouling part without biofouling interference.

10.1 Introduction

Biocide dosing [1], especially chock-dosing [2], have been proposed as an effective way to prevent biofouling in reverse osmosis and nanofiltration systems. This chapter aims to assess the effectiveness that a nonoxidizing biocide dosing such as DBNPA [3] might possess in preventing biofouling development. In this study, two different membranes have been used. The same feed spacer is selected in order to reduce as much as possible the different variables that can influence this research hypothesis.

Acknowledgments: The author would like to acknowledge the contributors of this chapter: Guillem Gilabert-Oriol, Gerard Massons, Tina Arrowood, Jon Johnson, and Steve Jons. The author would like to thank all members of the DuPont Water Solutions team, and specifically Nicolas Corgnet and Javier Dewisme for their outstanding help in performing these experiments.

This chapter is in the process of being submitted for publication, and was presented at the Society of Industrial Microbiology's Recent Advances in Microbial Control Meeting (RAMC) Conference in San Diego (USA) in October 2016.

https://doi.org/10.1515/9783111639079-010

10.2 Materials and methods

10.2.1 Wastewater characterization

Vila-seca wastewater treatment plant (WWTP) treats municipal wastewater from Vila-seca, La Pineda, and Salou villages [6]. The intake of the water is from the effluent of the clarifier, which is then pretreated using two DuPont™ Ultrafiltration SFP-2880 XP modules. The average ionic and organic characteristics of the ultrafiltered water used for the reverse osmosis experiments are summarized in Table 10.1.

Table 10.1: Wastewater feed characterization.

Parameter	Average concentration
ATP	0.03 ± 0.02 ng/L
TOC	8 ± 1.5 mg/L
Calcium	135 ± 16 mg/L
Chloride	509 ± 48 mg/L
Nitrate	63 ± 26 mg/L
Sodium	308 ± 30 mg/L
Sulfate	290 ± 49 mg/L
Conductivity	$2{,}408 \pm 442$ µS/cm
pH	7.3 ± 0.2

10.2.2 Experimental plant

The experimental plant used to evaluate the performance of the small 1812 membrane elements was the FREPP experimental plant, located in the WWTP in Vila-seca, Catalonia, Spain. A simple diagram and a photo of the asset can be found in Figure 10.1. This plant was used in order to perform the standard test during the soaking test. This plant was also used to perform the experiment with wastewater with a real DBNPA dosing. No sodium metabisulfite (SMBS) was used in any experiment so that its interference with DBNPA could be avoided. This experiment was performed at the Global Water Technology Center that DuPont Water Solutions has in Tarragona, Catalonia, Spain.

10.2.3 Membranes used

Smoother, more hydrophilic, and lower surface-charged membranes are described in the literature as being more fouling resistant [4–7]. Therefore, two 1812 membrane elements, membrane A and membrane B, were chosen in this study to represent examples

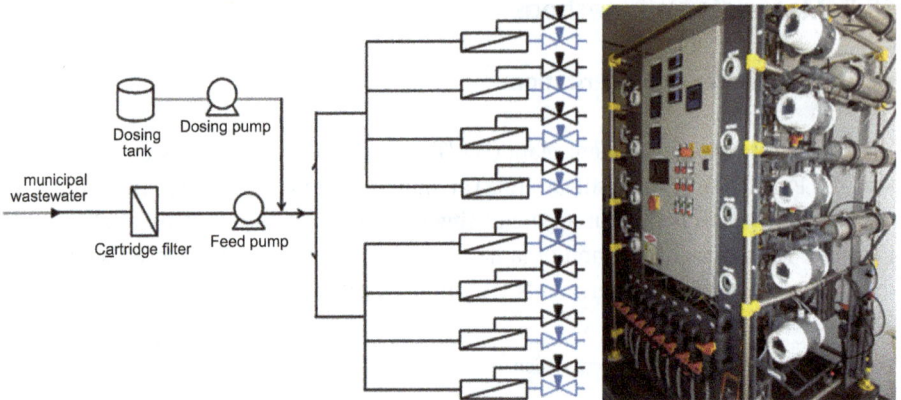

Figure 10.1: Diagram and picture of FREPP experimental plant.

of a low and high fouling resistant membrane surface, respectively. The properties of the two membranes are summarized in Table 10.2. Membranes A and B have similar degrees of roughness, but Membrane A is more hydrophobic and more negatively charged than Membrane B, and is expected to be fouled more rapidly. The values for the atomic force microscopy (AFM), the contact angle, and the zeta potential are displayed in the table below.

Table 10.2: Membrane characteristics.

Parameter	Membrane A	Membrane B
AFM average roughness	61 nm	61 nm
Contact Angle	72°	43°
Zeta potential (pH 8)	−25 mV	−5 mV

10.2.4 Operating conditions

Both 1812 membrane elements where first stabilized through a 5 min flushing using reverse osmosis permeate water that has a salinity of 1,000 ppm NaCl and 26 LMH of flux. This flushing was discarded and sent to the drain. Then, the stabilization protocol was adopted, followed by operating during 2 days in recirculation mode using a solution made of reverse osmosis permeate with 1,000 ppm NaCl and a flux of 26 LMH.

After this stabilization period, the membranes were operated at a constant flux of 18 LMH and at a constant recovery of 3.5%, dosing for 1 h/day with 20 ppm of biocide as is stated in the section below.

10.2.5 Dosing strategy

This experiment consisted of dosing AQUCAR™ DB 20 at 20 ppm DBNPA, 50 ppm CAR-BOWAX™ Polyethylene Glycol (PEG) 300, and 345 ppm CARBOWAX™ polyethylene glycol 200 for 1 h/day. This experiment used the FREPP experimental plant described in Figure 10.1, and it used real secondary effluent wastewater rafter being treated with DuPont ultrafiltration elements. This wastewater comes from the Vila-Seca WWTP that presents organic fouling potential. No sodium metabisulfite (SMBS) was used during this trial. In this experiment, two membrane chemistries are compared in the 1812 membrane element configuration. One uses membrane A and the other uses membrane B. Duplicates of both chemistries are used in order to better understand the signal-to-noise ratios between the groups. It is believed that biocide dosing will help mitigating biofouling, while PEG dosing will help accelerating organic fouling.

10.3 Results and discussion

10.3.1 Pressure drop

In the real dosing experiment, the normalized pressure drop evolution over time showed only a small biofouling increase during the first 5 months of operation, as seen in Figure 10.2. This fact highlights the benefit of shock-dosing DBNPA in order to mitigate biofouling in RO. The pressure drop values remain the same for membrane A and membrane B elements since the same feed spacer is in both elements. It is worth pointing out that while it seems that biocide shock-dosing is mitigating biofouling, biofouling still slowly develops over time, and after 5 months of operation, biofouling starts to develop normally. This might be explained as biocide has certainly an effect in eliminating bacteria, but once the biofilm is big enough, bacteria can shelter inside the EPS, and therefore biocide might stop being effective.

10.3.2 Salt rejection

It can also be observed that during these five months of shock-dosing of DBNPA, normalized salt rejection remained stable over the whole period, as shown in Figure 10.3. This might indicate that DBNPA dosing did not negatively affect the element performance. It should also be noticed that these elements were exposed to organic fouling, as shown in Figure 10.4. Membrane B elements offered higher salt rejection since it is a higher rejection membrane than the membrane A chemistry. The gap that exists in normalized salt rejection after month 3.5 was due to a failure in the permeate conduc-

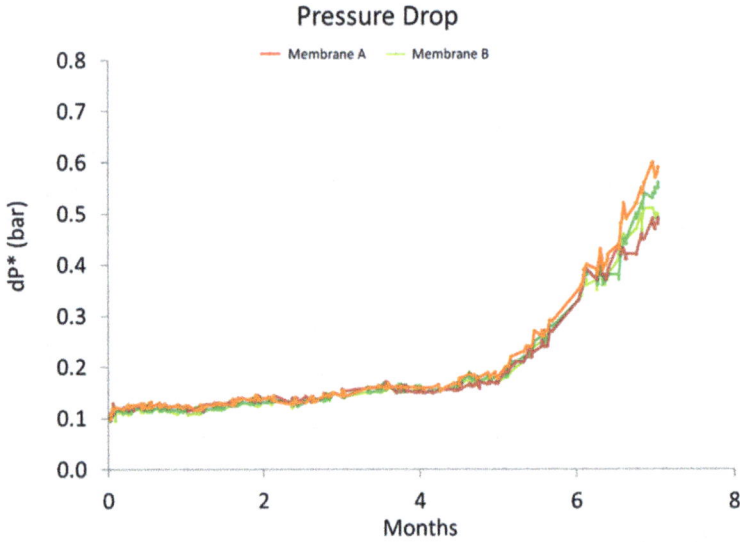

Figure 10.2: Normalized pressure drop evolution for Membrane A and Membrane B elements.

tivity transmitter. This was fixed after month 4. It should also be noticed the good reproducibility that the duplicates gave for each one of the two groups assessed. Finally, as the elements get more fouled, it can be seen how salt rejection tends to converge after more than 6 months of operation.

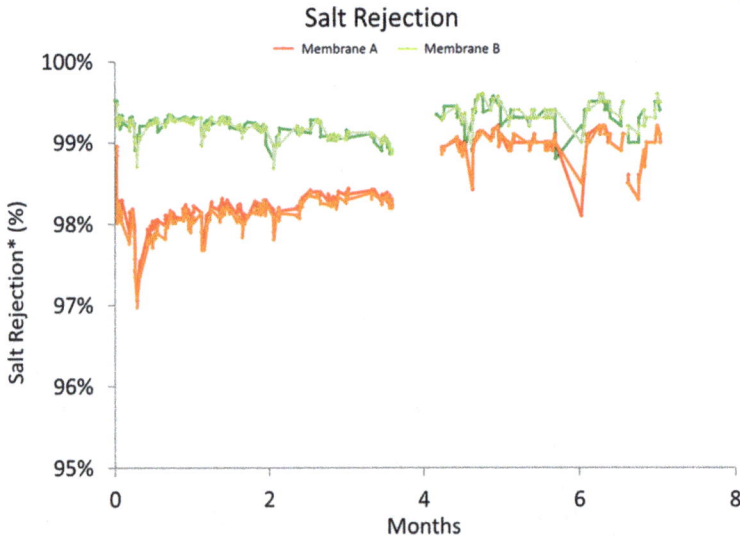

Figure 10.3: Normalized salt rejection evolution for Membrane A and Membrane B elements.

10.3.3 Water permeability

The water permeability, corresponding in this plot to the normalized A-value evolution, over time is plotted in Figure 10.4. In this graphic, it can be observed that water permeability decreased over time for both membrane A and membrane B elements due to organic fouling. It remains unclear if this observed fouling was due to the organics naturally present in the water with or without the influence of organic fouling coming from the PEG that the AQUCAR™ DB 20 has. Therefore, the next experiment studied the compatibility of DBNPA in a second pass RO, so that the water did not have any organic fouling coming from the water. It is also interesting to highlight that despite the membrane A element experiencing a faster decline in A-value over time during the 5 months experimental time, it was able to save energy compared to the membrane B elements. The good reproducibility that the duplicates gave for each one of the two groups assessed should also be noticed. Finally, as the elements get more fouled, it can also be seen how the difference in water permeability tends to become smaller after more than 6 months of operation. It should be noted that it was shown in a previous chapter that membrane B shows a better resistance to fouling, as it loses less permeability over time than membrane A.

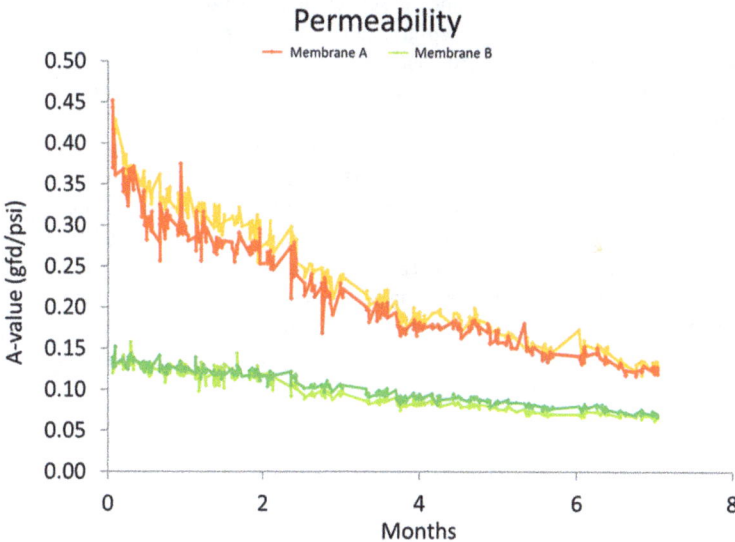

Figure 10.4: Normalized A-value evolution for membrane A and membrane B elements.

10.4 Conclusions

This research showed the impact of shock-dosing DBNPA, a nonoxidizing biocide in spiral-wound reverse osmosis membranes. It can be seen that while dosing the biocide initially helped in delaying and mitigating the biofouling development, after a certain time, biofouling still developed naturally. So, after 5 months of operation, biofouling starts to develop normally. This might be explained as biocide has certainly an effect in eliminating bacteria, but once the biofilm is big enough, bacteria can shelter inside the EPS, and therefore biocide might stop being effective. This study also showed how a fouling resistant membrane shows higher resilience over time, in terms of losing less water permeability over time. This also highlights the importance of having membrane chemistries that are fouling resistant. Dosing biocide is a useful method to study the first organic fouling part without biofouling interference.

References

[1]   L. H. Da-Silva-Correa, H. Smith, M. C. Thibodeau, B. Welsh, H. L. Buckley. The application of non-oxidizing biocides to prevent biofouling in reverse osmosis polyamide membrane systems: A review. AQUA – Water Infrastructure, Ecosystems and Society, 71(2) (2022) 261–292.
[2]   K. Majamaa, J. E. Johnson, U. Bertheas. Three steps to control biofouling in reverse osmosis systems. Desalination and Water Treatment, 42(1–3) (2012) 107–116.
[3]   G. R. Ras. Control of Biofouling on Reverse Osmosis Membranes Using DBNPA, Stellenbosch University. (2016).
[4]   M. Herzberg, M. Elimelech. Biofouling of reverse osmosis membranes: Role of biofilm enhanced osmotic pressure. Journal of Membrane Sciences, 295 (2007) 11–20.
[5]   M. Herzberg, S. Kang, M. Elimelech. Role of extracellular polymeric substances (EPS) in biofouling of reverse osmosis membranes. Environmental Science & Technology, 43 (2009) 4393–4398.
[6]   E. M. Van Wagner, A. C. Sagle, M. M. Sharma, Y. H. La, B. D. Freeman. Surface modification of commercial polyamide desalination membranes using poly (ethylene glycol) diglycidyl ether to enhance membrane fouling resistance. Journal of Membrane Science, 367 (2011) 273–287.
[7]   E. Matthiasson. The role of macromolecular adsorption in fouling of ultrafiltration membrane. Journal of Membrane Science, 16 (1983) 23–36.

Chapter 11
Visualizing biofouling on the membrane surface

Development of Fortilife™ Director™, a novel method for biofilm visualization on reverse osmosis membranes

Imaging techniques are an important research tool for the qualitative study of static biofilms. Biofilm samples should ideally be visualized with minimal sample preparation to not alter its original structure. However, this can be challenging, as these compounds have little to no color and are highly hydrated. Traditional image-based techniques such as confocal laser scanning microscopy (CLSM) or optical coherence tomography (OCT) are expensive, challenging, and labor-intensive. This chapter details the development of a novel method to visualize in situ biofouling on a solid surface. Fortilife™ Director™ method focuses on covering the fouled membrane surface with a special reagent, creating a homogeneous and thin layer that reveals areas where biofouling is deposited. Finally, the surface covered is quantified through image analysis software. This method offers a high contrast of the boundaries of biofilm and its morphology. This allows determining the percentage of membrane surface covered by biofilm. Images of biofilm at different growing phases and calculations of biofilm impact based on surface area are provided. Finally, the results of the method are compared against traditional analytical methods for biofilm quantification. Biofilm coverage is found to be more correlated with the reported operational effects of biofouling than any other method.

11.1 Introduction

11.1.1 Biofouling on reverse osmosis membranes

One of the main hurdles for membrane filtration using challenging feed water is biofouling [1, 2]. Biofouling is the growth of biofilm within the module that causes flux decline and an increase on energy consumption and chemical cleaning frequency, lowering the membrane lifetime [3]. Biofilm is defined as a complex assemblage of

Acknowledgments: The author would like to acknowledge the inventor of this methodology, Gerard Massons, as well as the contributors of this chapter: Gerard Massons, Guillem Gilabert-Oriol, Can Wu, Veronica Gomez, Tina Arrowood, and Veronica Garcia Molina. The author would also like to thank all members of DuPont Water Solutions team for their support on doing this research.

This chapter is in the process of being submitted for publication, and it has been published as a preprint with the following reference: G. Massons Gassol, G. Gilabert-Oriol, V. Gomez, T. Arrowood, V. Garcia Molina, Fortilife™ Director™ Biofouling Diagnostic Tool Developed for Biofilm Visualization on Reverse Osmosis Membranes, SSRN (2024).

https://doi.org/10.1515/9783111639079-011

microbial communities attached to a substratum and to each other by means of a dense, self-produced biopolymer matrix of extracellular polymeric substances (EPSs). This hydrogel is mainly composed of polysaccharides and proteins and has high-water content [4]. EPS has been recognized as the main contributor to the decline in operational performance of elements suffering from biofouling [3, 5]. A comparable effect is caused by transparent exopolymer particles (TEP), which share similar characteristics with EPS [6]. TEP had recently raised awareness as it is commonly found during algal blooms and causes severe decrease on membrane permeability [7].

The main research focus to improve biofouling resistance on reverse osmosis (RO) operation relies on the optimization of the feed spacer [2, 8, 9], membrane chemistry [10, 11], and pretreatment [12]. Visual inspection remains an essential tool to study biofilm formation on the membrane surface and evaluate the impact of feed spacer and membrane chemistry modifications [13, 14]. Thus, imagining techniques are an important research tool for the qualitative study of biofilms. Biofilm visualization is used for a fundamental understanding of the relationship of hydrodynamics and operating conditions on biofouling development, accumulation, and distribution [15, 16]. Biofilm samples should ideally be visualized with minimal sample preparation to not alter its original distribution, structure, dimensions, and thickness [17]. However, visualization of biofouling samples is challenging, as biopolymers have little or no color and are highly hydrated. Different techniques have been developed over the last years. However, most of them are time-consuming, require biofilm sample manipulation, or only provide good definition at a microscopic resolution, and cannot be applied on large areas.

11.1.2 State-of-the-art biofouling visualization techniques

One of the most common techniques to examine biofilms in the microscale is CLSM. Fluorophore dyes that target specific domains within the biofilm are used to stain the samples [18]. However, the main disadvantages are the high cost of confocal microscope systems and that protocols are labor intensive, as fluorophores are often unlikely to stain all biofilm components [19]. Additionally, reverse osmosis and nanofiltration membranes generate an autofluorescent background when visualized using CLSM, which creates interferences, making it difficult to distinguish the biofilm from the membrane surface [20].

Determining biofilm distribution using high-resolution techniques can be misleading due to biofouling heterogeneity. Mesoscale (mm-scale) methods allow larger area of several millimeters to be inspected and obtain a more representative picture of the large-scale effects [18]. Examples of methods successfully applied to study biofilm growth on RO membranes include OCT [21] and magnetic resonance imaging (MRI) [16]. OCT records a thin optical slice of the sample based on the scattering properties of biofilm, using a light source and photodetectors without the need for staining [22]. MRI biofilm imaging is based on the differences in fluid velocity on the bulk phase, revealing the biofilm boundaries as a reduction on the crossflow velocity [23]. Both methods are based on indirect measurements of the changes on refraction index or fluid velocity caused by the biofilm structures. The main drawbacks of these methods are interferences and/or background noise in the measurements, the limited resolution, and the high cost of these systems [19].

Biofilm structures can also be revealed by applying dyes that are color-sensitive to oxidation–reduction reactions, such as aerobic respiration. Different oxygen consumption rates reveal the areas with viable biofilm. Most used dyes as redox indicators are resazurin [24] and 5-cyano-2,3-ditolyl tetrazolium chloride (CTC) dye [2]. However, detecting areas with biofilm presence by changes in oxygen concentration is challenging, as oxygen and the dyes are diffusing on the liquid and biofilm edges and become blurry. Additionally, bacteria metabolic activity can be affected by external conditions and does not correlate with EPS concentration [25], and colored areas present blurry edges, as dyes diffuses on the liquid phase.

Biofilm components can also be stained to discover the areas where EPS are attached to the surface, using what is known as positive stain. The principle is that dyes, such as crystal violet [26] or Coomassie blue [27], will diffuse and stain the biofilm matrix when applied. The excess dye is then rinsed, leaving the EPS-related components colored. The critical steps for these methods are good diffusion of the dye through the biofilm matrix, efficient rinsing without altering the biofilm structure, and removing the dye from the clean membrane surface. This last step is critical to obtain good contrast between the background membrane and the biofilm structure but can be challenging when studying biofilm on polyamide RO membranes due to the chemical similarity to proteins.

Negative stain methods can also be used to visualize biofilm, using a dye that is able to diffuse in the liquid phase, but not in the EPS matrix. Rhodamine is one dye used as a negative stain. It can be injected during operation to reveal flow patterns created by biofilm growth [28, 29]. The free moving water areas are where rhodamine can diffuse, as biofilm is not present. However, visualization is only feasible for extreme biofouling scenarios (preferential flow paths present) and only short after rhodamine is injected, as it quickly diffuses into the biofilm matrix.

11.1.3 Fortilife™ Director™ biofilm visualization tool

A method using microparticles suspension for in situ visualization of biofilm structures on solid surfaces, specifically designed for membrane surface, is described herein. This method is herein referred as Fortilife™ Director™ biofilm visualization tool.

This novel biofouling visualization method overcomes some of the shortcomings from the existing state of the art methods. It can be applied to the attached biofilm without altering its morphology. Minimal equipment is needed, and a high contrast of the biofilm's boundaries can be observed. This method allows for a fundamental understanding of biofilm development and distribution under different conditions and can quantify the surface area impacted by biofouling. Although the examples provided are from biofouled RO membrane elements, it can be used in a variety of membranes, including reverse osmosis, nanofiltration, ultrafiltration, or microfiltration.

11.2 Materials and methods

11.2.1 Fortilife™ Director™ overview

For surface biofouling visualization, the liquid used contains small, uniformly sized particles so that it can spread evenly on the membrane surface. The particles contained in the Fortilife™ Director™ reagent are carbon microparticles and have a range of 0.01–2 μm, with a concentration in the liquid suspension of 0.5–10 w/v% [30]. Such microparticles have a very low diffusion coefficient into the EPS matrix, thus preventing it from penetrating the biofilm structure [31]. Thus, the background is darkened while biofilm remains clear. The protocol can be directly applied in situ, with minimal sample preparation, to prevent altering the biofilm morphology. Fortilife™ Director™ offers a sharp contrast of the boundaries of the biofilm, not found in most other biofilm visualization procedures. The areas with biofouling are not affected and as a result, appear as bright spots when applying backlight. The edge of the biofilm is highlighted with crispiness, like in a contour plot, as schematically represented in Figure 11.1.

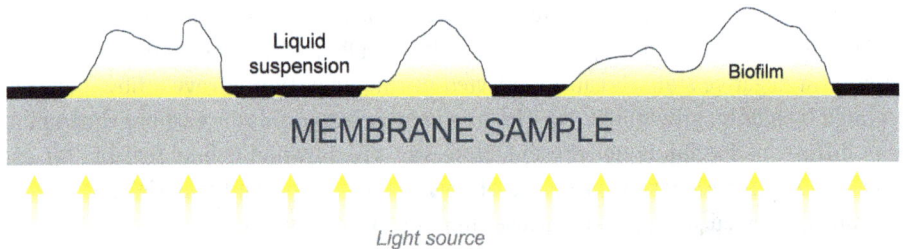

Figure 11.1: Principle of Fortilife™ Director™ for biofouling visualization on membrane sample.

11.2.2 Comparison to alternative biofilm visualization procedures

The reagent to be applied on the RO samples for biofilm staining should only interact with the biofilm to provide good discernment from background (membrane surface). Different compounds commonly used to visualize biofilm, for either positive or negative staining, are tested to evaluate their compatibility with virgin RO membrane surface. The stains used should not stain the active layer of the clean membranes once rinsed, as the remaining background color will create a poor contrast with the biofilm sample. To validate compatibility, membrane coupons are soaked in crystal violet, lugol iodine, safranine, rhodamine B, and the microparticle suspension of Fortilife™ Director™. After 3 min of contact time, all coupons are rinsed with ultrapure water to determine any permanent staining of the polyamide layer.

11.2.3 Biofilm visualization

Once the liquid suspension is applied, the biofouled samples can be visualized with naked eye or with the aid of a stereomicroscope, in this case, a Leica MS5 is used. The stereomicroscope magnification renders from 0.63× to 4×, allowing for realistic visualization of membrane elements, such as fouled feed spacer or membrane. The instrument uses two separate optical paths with two objectives and eyepieces. The result produces a three-dimensional visualization of the sample. Back illumination is used, as it increases the contrast between the dark background and the areas where biofouling is present, which appears as clear and bright spots.

To obtain a picture for documentation and further analysis, a 12.1-megapixel digital camera, Canon Digital Ixus 200 IS, is used. The area of the membrane captured with the camera can vary from 7.6 to 510.7 mm^2, depending upon the magnification level used in the stereomicroscope.

11.2.4 Biofilm surface coverage calculation

The stages of biofilm development and the extent of the formed biofouling over the membrane's coupon can be studied in detail using the images generated in the method herein described. Biofilm surface area coverage is estimated from the membrane sample pictures acquired using the open-source image analysis software, Image J 1.41 [34]. This software has been used earlier to process biofilm imaging results obtained with different techniques [35–37].

Acquired images are subsequently processed. The first step is to convert the 3-channel (red, green, and blue [RGB]) picture to an 8-bit grayscale image. The study of the intensity distribution diagram of the RGB channels showed that splitting channels and using the green channel offered the best results in order to convert the original RGB membrane sample pictures to 8-bit grayscale.

To achieve a reliable binarization, the resulting image is edited with a black-and-white threshold filter, adjusted to exclude the membrane surface, which is much darker than the areas with biofilm. The threshold value is determined by applying the protocol to a clean membrane and adjusting the filter to exclude all background pixels. The number of pixels above a defined threshold intensity divided by the total number of pixels corresponds to area covered by biofilm on a per-pixel basis. The biofilm surface coverage (CV) is calculated by counting only the biofilm signal pixels (P_w, intensity = 255) and divided by the total pixel count (P_T), according to eq. (11.1). It should be noticed that when no biofouling is present, the measured CV is expected to be 0%, whereas a CV of 100% means that the membrane surface is fully covered by biofilm slime.

Biofilm surface coverage:

$$CV = \frac{P_w}{P_T} \qquad\qquad (11.1)$$

11.2.5 Biofouling quantification in different fouling matrices

The biofouling samples used for validating this method are obtained from different types of RO samples. Sampling is always performed in the mid-region of the membrane leaf or coupon, to obtain a representative distribution of the feed-concentrate. All these experiments were performed in the Global Water Technology Center that DuPont Water Solutions has in Tarragona, Catalonia, Spain.

One set of samples were obtained from coupons operated in a membrane fouling simulator (MFS) flat cell unit [32]. Operational data in terms of pressure drop increase is contrasted with the amount of biofouling present in the coupons after operation using brackish water [33].

Additional samples of 4-inch membrane elements and 2.5-inch membrane elements operated in a two-stage system using municipal wastewater (62% recovery) are analyzed. Elements were autopsied after 6-month operation to demonstrate biofouling distribution in a staged system, with multiple elements in series and under a high organic load scenario.

Finally, the results of the tail 8-inch membrane element operated in a two-stage system using city water (75% recovery) are also provided. The element was autopsied after 8-month operation to evaluate the fouling present that was causing high pressure drop and permeate conductivity.

Recovery	Element type	Size	Most predominant fouling types	Water parameters
0%	Flat cell	20 × 4 cm	Biofouling and organic	TDS: 1,000 mg/L HCO_3: 160 mg/L TOC: 1.3 Temperature: 19 °C
62%	Spiral wound	4-inch (stage 1) 2.5-inch (stage 2)	Organic fouling, biofouling, and scaling	TDS: 1,500 mg/L HCO_3: 400 mg/L TOC: 6.9 Temperature: 22 °C

(continued)

Recovery	Element type	Size	Most predominant fouling types	Water parameters
75%	Spiral wound	8-inch	Scaling, organic, and biofouling	TDS: 400 mg/L HCO_3: 150 mg/L TOC: 1.0 Temperature: 10 °C

11.3 Results and discussion

11.3.1 Comparison to alternative biofilm visualization procedures

A representative selection of traditional microbiological stains is applied on the polyamide side of clean RO membrane coupons. These stains are rhodamine, safranine, lugol, and crystal violet. These are compared with the Fortilife™ Director™ biofilm visualization tool, as well as with a blank, consisting of just water.

After 3 min, it is rinsed with ultrapure water until clear to remove excess dye. The color on the area in contact with the different dyes is compared to a virgin membrane soaked only in water.

As can be observed in Figure 11.2, all compounds permanently stained membranes, even after rinsing. Only the coupon soaked using Fortilife™ Director™ method is efficiently rinsed with water, because the particles contained in the reagent do not permanently attach to the membrane surface. This test confirms the technical difficulties of visualizing biofilm on membrane samples, since traditional dyes described in the literature stain both the polyamide and the biofilm, limiting its usefulness.

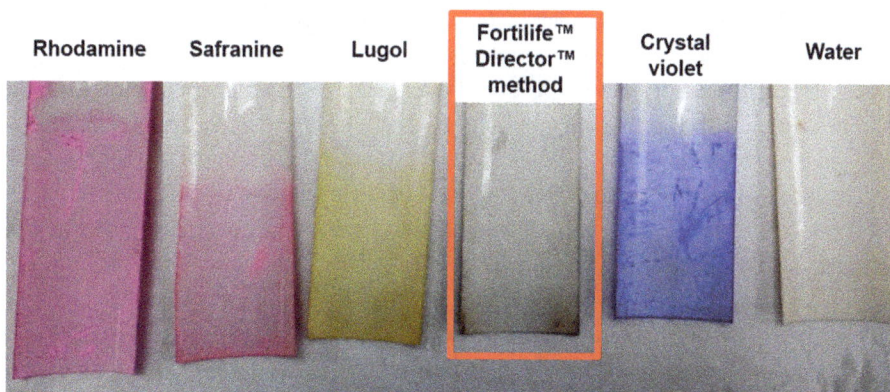

Figure 11.2: Color remaining in the polyamide side after rinsing.

11.3.2 Biofilm visualization example

When a thin layer of biofouling is present in the RO membrane, it is very difficult to visually characterize the amount or its distribution on the membrane surface. However, by using Fortilife™ Director™ method on the sample, the areas with biofouling are revealed. An example showing the same RO sample before and after applying the method can be observed in Figure 11.3. The true impact of biofouling can only be assessed after applying Fortilife™ Director™ in the sample. Once the method is applied, the amount and growth pattern of biofilm on the membrane sample can be easily determined.

Membrane before applying Fortilife™ Director™   **Membrane after applying Fortilife™ Director™**

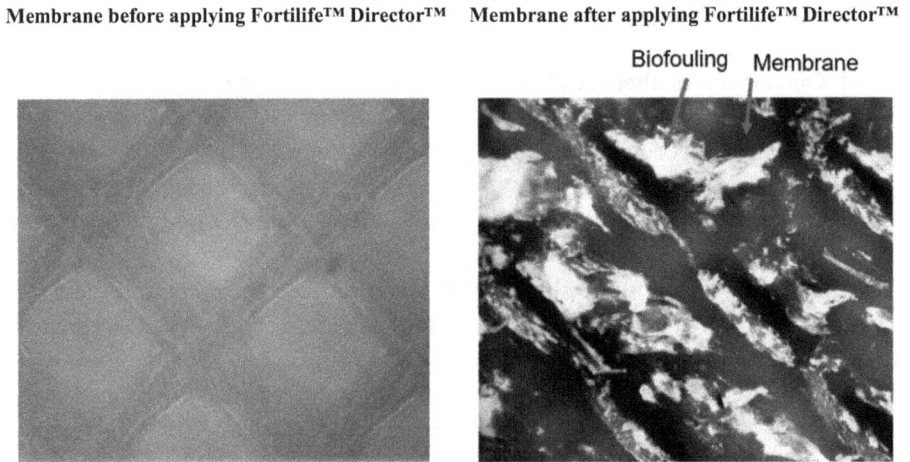

Figure 11.3: RO membrane sample with biofouling before (left) and after (right) applying Fortilife™ Director™.

The images obtained when using a stereomicroscope offer a great level of detail at different scales (from millimeters to micrometers), as can be observed in Figure 11.4, with the raw images. Microbial biofilms can be microscopically resolved from large to small magnification levels, depending on the objective of the research. Because of the heterogeneity of biofilm structures, the sample might need to be examined over a large area [38, 39]. However, other research might be focused on studying small details on the biofilm. One big advantage of Fortilife™ Director™ is that it offers flexibility for clear visualization at different magnification levels: from general membrane surface to biofilm microstructures.

11.3.3 Biofilm surface coverage quantification

The amount of biofouling present in the membrane surface can be measured and quantified using image refinement and transformation. The percentage of white pixels can be associated with the biofilm surface coverage percentage, as areas with biofilm are

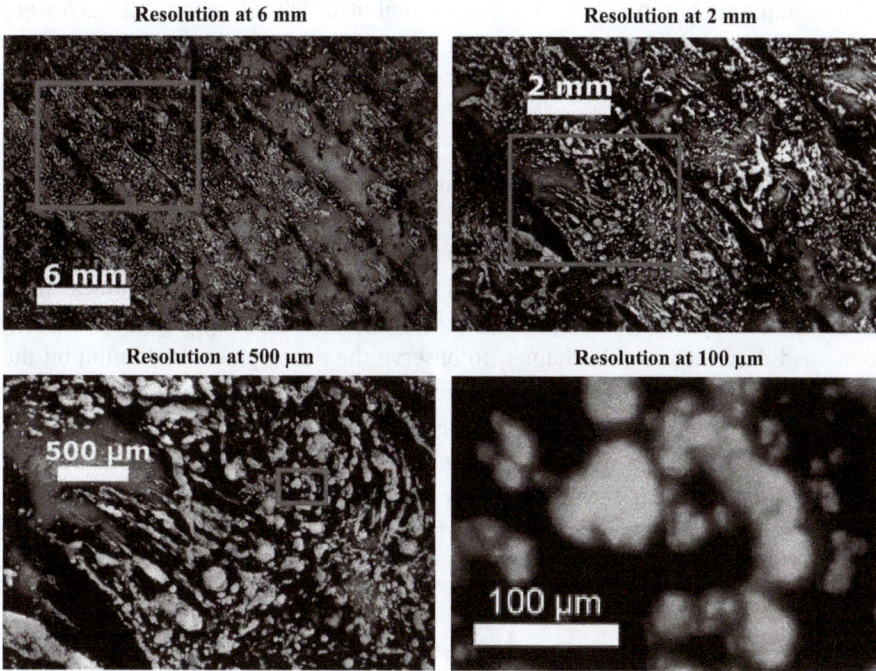

Figure 11.4: Detail of the resolution obtained for macro- and micro-scale visualization for raw images.

highlighted when applying backlight. An example of the biofilm surface coverage quantification is presented in Figure 11.5. The original picture was transformed to a black-and-white image, adjusting the threshold value to exclude all gray tones corresponding to the background membrane surface. The result is a binary picture, where the percentage of white pixels corresponds to the surface area colonized by the biofilm. As shown in Figure 11.1, only biofilm structures with a minimum relevant thickness are highlighted. Additionally, since biofilm thickness could be limited by shear stress during operation [16, 21, 29], colonized surface area can be a representative parameter to monitor the different stages of biofilm development and their impact on operational performance.

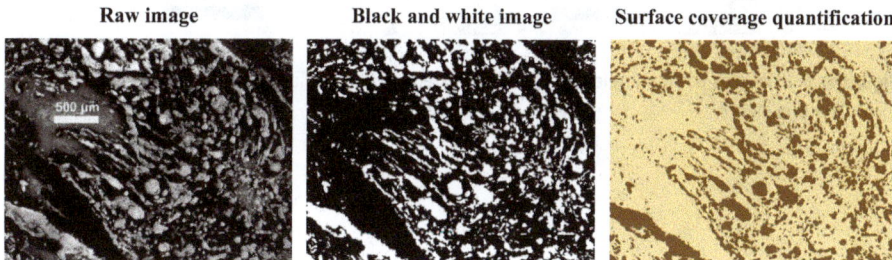

Figure 11.5: Conversion of raw image to black-and-white to quantify biofilm surface coverage.

In the example shown in Figure 11.5, from a total of 12,000,000 pixels, 3,255,286 were areas with biofouling. The surface coverage is calculated using eq. (11.1), showing that surface coverage is 27%.

11.3.4 Biofouling quantification in different fouling matrices

11.3.4.1 Flat-sheet coupons operated with river water

The fouling severity can be rapidly assessed using Fortilife™ Director™. As an example, four different samples containing increasing amounts of biofouling are analyzed using the method described in this chapter, to observe the development of biofilm on the membrane surface. All repetitions were operated in the same conditions but were analyzed at different equally spaced times (3 days), to track the evolution of biofouling development. From the biofilm distribution shown in Figure 11.6 it can be confirmed that feed spacer is the starting seeding point for biofilm development (sample A) and how rapidly biofouling takes over membrane surface once it is colonized (sample B < C≪D).

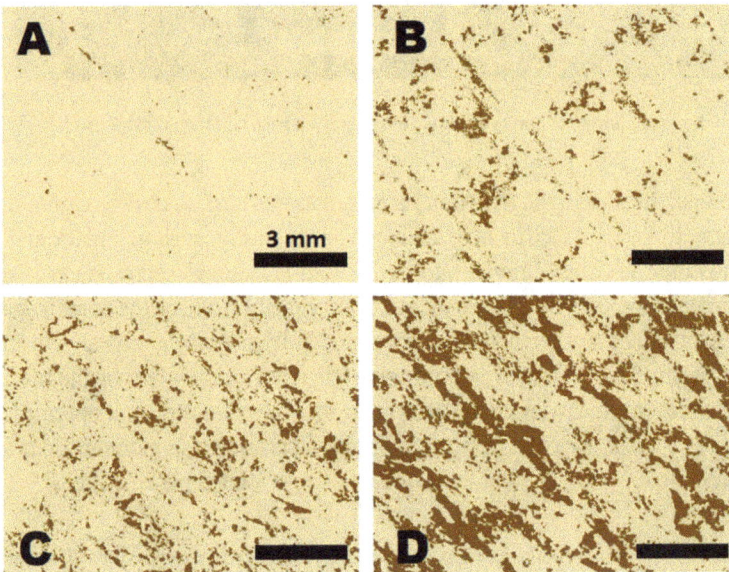

Figure 11.6: Examples of RO samples having low (A), medium (B), medium-high (C), and high (D) biofouling prevalence.

Surface coverage was calculated for each of the samples. Surface coverage was compared to other typically used parameters to quantify biofouling [40] such as adenosine triphosphate (ATP), total organic carbon (TOC), total nitrogen (TN), proteins (PN), and carbohydrates (CH). The operational impact of biofouling in terms of pressure drop increase is presented in Table 11.1 together with analytical results.

Table 11.1: Autopsy results of RO samples showing increasing levels of biofouling.

Parameter	Figure 11.6a	Figure 11.6b	Figure 11.6c	Figure 11.6d
Operating time (days)	3	6	9	12
dP increase (%)	17%	52%	193%	650%
Surface coverage (%)	1%	7%	12%	32%
ATP (ng/cm^2)	3.6	0.1	5.0	112
TOC (mg/m^2)	34	17	218	522
TN (mg/m^2)	16	13	105	169
PN (mg/m^2 N)	0	31	227	484
CH (mg/m^2 C)	0	0	59	506

As can be observed in Table 11.2, surface coverage showed a statistically significant correlation with pressure drop increase. Among the different parameters, biofilm coverage quantified showed the highest correlation and best *p*-value to pressure drop increase.

Table 11.2: Statistical analysis compared to pressure drop increase.

Parameter	Correlation (R^2)	p-Value
Surface coverage	0.991	0.009
ATP	0.970	0.030
TOC	0.987	0.013
TN	0.940	0.060
PN	0.979	0.021

Table 11.3 shows that surface coverage is also highly correlated to the traditional analytical parameters used for biofilm quantification. These overall results confirm that Fortilife™ Director™ is more reliable and representative to be used for diagnosis than alternative analytical methods.

11.3.4.2 Spiral-wound elements operated with municipal wastewater

During testing time summarized in Figure 11.7 a much steeper increase in the pressure drop of the elements from the first staged compared to the elements of the second stage can be observed. It also shows that a gradual larger portion of pressure drop increase cannot be fully removed after each chemical cleaning.

Table 11.3: Statistical analysis compared to surface coverage.

Parameter	Correlation (R^2)	p-Value
ATP	0.945	0.055
TOC	0.972	0.028
TN	0.933	0.067
PN	0.977	0.023
CH	0.968	0.032

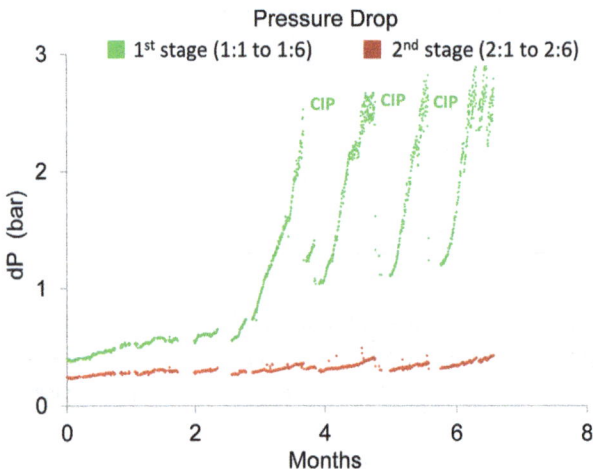

Figure 11.7: Pressure drop distribution in a two-staged system.

Three elements of each stage were autopsied after operation and after being chemically cleaned three times. The objective is to determine biofouling distribution by using Fortilife™ Director™ on elements from different positions within a pressure vessel. Figure 11.8 represents the average surface coverage determined for each element position in a pressure vessel. A much higher prevalence of biofouling in the first element from first stage can be observed. Although pressure drop measurement for first stage measures the differential pressure of six elements in series, it can be attributed mainly to first element. This test also shows no influence when analyzing elements from a staged system with inorganic and organic foulants.

Figure 11.9 is presented as a visual aid for the biofouling distribution observed on the membrane elements. The first elements showed more dense and aggregated biofilm structures that are becoming more evenly spread into small colonies in the tail elements. This could be attributed to lower shear present in the last elements combined with the entrapment and clogging typically observed in first elements.

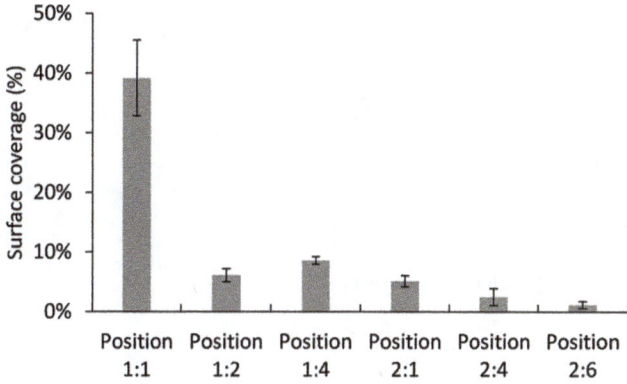

Figure 11.8: Biofouling distribution in a two-staged system.

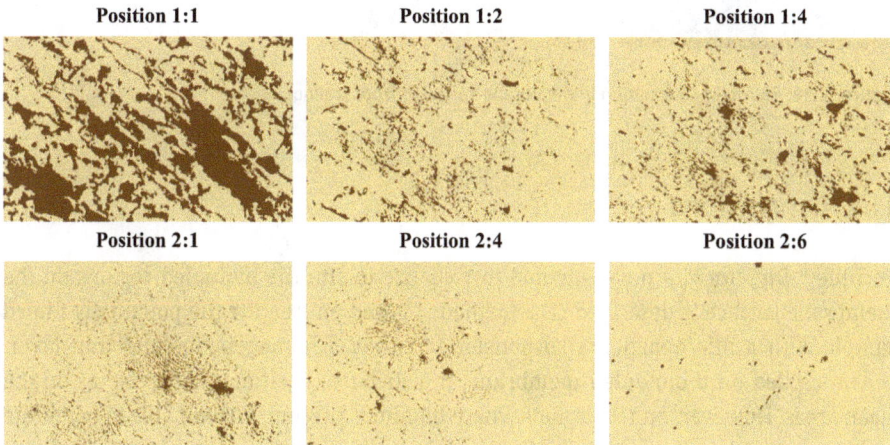

Figure 11.9: Biofouling distribution in a two-staged system.

11.3.4.3 Spiral-wound elements operated with city water

The performance of the tail element showed very high pressure drop of 2.9 bar and high permeate conductivity. Once autopsied severe scaling, mainly from calcium carbonate, could be observed . These samples were used to show that there is no interference of inorganic components accumulated on the membrane surface with the method. Additionally, the results in Figure 11.10 can be used to confirm that visualization of potential areas with biofouling is not impacted even when severe inorganic scaling is present.

Original sample

Biofouling location

Original sample (zoomed)

Biofouling location (zoomed)

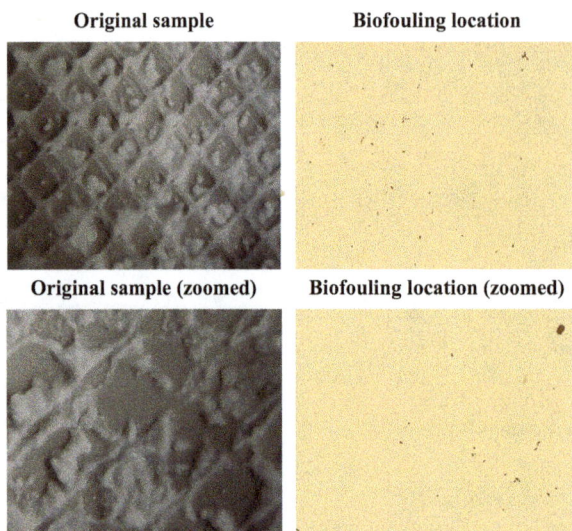

Figure 11.10: Biofouling evaluation in membrane suffering from scaling.

11.4 Conclusions

Fortilife™ Director™, a novel method to visualize in situ the biofouled regions on the membrane surface is described. The method is based on treating the potentially fouled sample with a microparticle's suspension, creating a homogeneous and thin layer. When applied on a biofouled membrane, it distributes itself and accumulates on the clean areas. However, on the regions where biofilm is present, it is not able to penetrate and thus, reveals areas where biofouling was deposited when using backlight. Fortilife™ Director™ offers a high contrast of the boundaries of biofilm. Also, morphology is not altered, as the protocol is applied directly on the membrane surface.

This method offers a detailed determination of biofouling distribution. Quantitative results can be obtained after processing the images obtained with this method, allowing the biofilm surface coverage to be calculated. It was demonstrated that the biofilm surface coverage results have high correlation with fouling-associated operational performance data (pressure drop increase). It also proves to be more reliable and representative to be used for diagnosis than alternative fouling quantification methods traditionally used when other types of fouling is present.

References

[1]   Y. N. Wang, C. Y. Tang. Fouling of nanofiltration, reverse osmosis, and ultrafiltration membranes by protein mixtures: The role of inter-foulant-species interaction. Environmental Science & Technology, 45(15) (2011) 6373–6379.

[2]   S. R. Suwarno, X. Chen, T. H. Chong, V. L. Puspitasari, D. McDougald, Y. Cohen, S. A. Rice, A. G. Fane. The impact of flux and spacers on biofilm development on reverse osmosis membranes. Journal of Membrane Science, 405 (2012) 219–232.

[3]   M. Jafari, A. D'haese, J. Zlopasa, E. R. Cornelissen, J. S. Vrouwenvelder, K. Verbeken, C. Picioreanu. A comparison between chemical cleaning efficiency in lab-scale and full-scale reverse osmosis membranes: Role of extracellular polymeric substances (EPS). Journal of Membrane Science, 609 (2020) 118189.

[4]   C. Dreszer, J. S. Vrouwenvelder, A. H. Paulitsch-Fuchs, A. Zwijnenburg, J. C. Kruithof, H. C. Flemming. Hydraulic resistance of biofilms. Journal of Membrane Science, 429 (2013) 436–447.

[5]   T. Yu, H. Sun, Z. Chen, Y. H. Wang, Z. Y. Huo, N. Ikuno, K. Ishii, Y. Jin, H. Y. Hu, Y. H. Wu, Y. Lu. Different bacterial species and their extracellular polymeric substances (EPSs) significantly affected reverse osmosis (RO) membrane fouling potentials in wastewater reclamation. Science of the Total Environment, 644 (2018) 486–493.

[6]   N. Saeidi, M. Rzechowicz, M. F. Siddiqui, S. W. Teng, A. G. Fane, H. Winters. The potential role of transparent exopolymer particles (TEP) and their pseudo-precursors on biofouling of RO water treatment membranes. Desalination and Water Treatment, 182 (2020) 1–12.

[7]   S. Meng, R. Wang, K. Zhang, X. Meng, W. Xue, H. Liu, D. Liang, Q. Zhao, Y. Liu. Transparent exopolymer particles (TEPs)-associated protobiofilm: A neglected contributor to biofouling during membrane filtration. Frontiers of Environmental Science & Engineering, 15(4) (2021) 1–10.

[8]   K. Reid, M. Dixon, C. Pelekani, K. Jarvis, M. Willis, Y. Yu. Biofouling control by hydrophilic surface modification of polypropylene feed spacers by plasma polymerization. Desalination, 335(1) (2014) 108–118.

[9]   R. V. Linares, S. S. Bucs, Z. Li, M. AbuGhdeeb, G. Amy, J. S. Vrouwenvelder. Impact of spacer thickness on biofouling in forward osmosis. Water Research, 57 (2014) 223–233.

[10] M. Ben-Sasson, X. Lu, E. Bar-Zeev, K. R. Zodrow, S. Nejati, G. Qi, M. Elimelech. In situ formation of silver nanoparticles on thin-film composite reverse osmosis membranes for biofouling mitigation. Water Research, 62 (2014) 260–270.

[11] D. Saeki, S. Nagao, I. Sawada, Y. Ohmukai, T. Maruyama, H. Matsuyama. Development of antibacterial polyamide reverse osmosis membrane modified with a covalently immobilized enzyme. Journal of Membrane Science, 428 (2013) 403–409.

[12] K. J. Varin, N. H. Lin, Y. Cohen. Biofouling and cleaning effectiveness of surface nanostructured reverse osmosis membranes. Journal of Membrane Science, 446 (2013) 472–481.

[13] C. Sun, L. Fiksdal, A. Hanssen-Bauer, M. B. Rye, T. Leiknes. Characterization of membrane biofouling at different operating conditions (flux) in drinking water treatment using confocal laser scanning microscopy (CLSM) and image analysis. Journal of Membrane Science, 382(1–2) (2011) 194–201.

[14] P. Xu, C. Bellona, J. E. Drewes. Fouling of nanofiltration and reverse osmosis membranes during municipal wastewater reclamation: Membrane autopsy results from pilot-scale investigations. Journal of Membrane Science, 353(1–2) (2010) 111–121.

[15] L. Vanysacker, R. Bernshtein, I. F. Vankelecom. Effect of chemical cleaning and membrane aging on membrane biofouling using model organisms with increasing complexity. Journal of Membrane Science, 457 (2014) 19–28.

[16] S. A. Creber, T. R. R. Pintelon, D. G. Von der Schulenburg, J. S. Vrouwenvelder, M. C. M. Van Loosdrecht, M. L. Johns. Magnetic resonance imaging and 3D simulation studies of biofilm

accumulation and cleaning on reverse osmosis membranes. Food and Bioproducts Processing, 88(4) (2010) 401–408.

[17] E. Bar-Zeev, K. R. Zodrow, S. E. Kwan, M. Elimelech. The importance of microscopic characterization of membrane biofilms in an unconfined environment. Desalination, 348 (2014) 8–15.

[18] T. R. Neu, J. R. Lawrence. Innovative techniques, sensors, and approaches for imaging biofilms at different scales. Trends in Microbiology, 23(4) (2015) 233–242.

[19] J. R. Lawrence, G. D. W. Swerhone, G. G. Leppard, T. Araki, X. Zhang, M. M. West, A. P. Hitchcock. Scanning transmission X-ray, laser scanning, and transmission electron microscopy mapping of the exopolymeric matrix of microbial biofilms. Applied and Environmental Microbiology, 69(9) (2003) 5543–5554.

[20] A. J. Semião, O. Habimana, E. Casey. Bacterial adhesion onto nanofiltration and reverse osmosis membranes: Effect of permeate flux. Water Research, 63 (2014) 296–305.

[21] C. Dreszer, A. D. Wexler, S. Drusová, T. Overdijk, A. Zwijnenburg, H. C. Flemming, J. C. Kruithof, J. S. Vrouwenvelder. In-situ biofilm characterization in membrane systems using Optical Coherence Tomography: Formation, structure, detachment and impact of flux change. Water Research, 67 (2014) 243–254.

[22] S. West, M. Wagner, C. Engelke, H. Horn. Optical coherence tomography for the in situ three-dimensional visualization and quantification of feed spacer channel fouling in reverse osmosis membrane modules. Journal of Membrane Science, 498 (2016) 345–352.

[23] T. R. Neu, B. Manz, F. Volke, J. J. Dynes, A. P. Hitchcock, J. R. Lawrence. Advanced imaging techniques for assessment of structure, composition and function in biofilm systems. FEMS Microbiology Ecology, 72(1) (2010) 1–21.

[24] M. E. Sandberg, D. Schellmann, G. Brunhofer, T. Erker, I. Busygin, R. Leino, P. M. Vuorela, A. Fallarero. Pros and cons of using resazurin staining for quantification of viable Staphylococcus aureus biofilms in a screening assay. Journal of Microbiological Methods, 78(1) (2009) 104–106.

[25] M. L. Yallop, D. M. Paterson, P. Wellsbury. Interrelationships between rates of microbial production, exopolymer production, microbial biomass, and sediment stability in biofilms of intertidal sediments. Microbial Ecology, 39(2) (2000) 116–127.

[26] E. Peeters, H. J. Nelis, T. Coenye. Comparison of multiple methods for quantification of microbial biofilms grown in microtiter plates. Journal of Microbiological Methods, 72(2) (2008) 157–165.

[27] G. Boels, G. Blackman, A. Calabozo, Kit for detecting biofilms, U.S. Patent 20140113326, (2024)

[28] E. I. Prest, M. Staal, M. Kühl, M. C. Van Loosdrecht, J. C. Vrouwenvelder. Quantitative measurement and visualization of biofilm O2 consumption rates in membrane filtration systems. Journal of Membrane Science, 392 (2012) 66–75.

[29] R. Valladares Linares, L. Fortunato, N. M. Farhat, S. S. Bucs, M. Staal, E. O. Fridjonsson, T. Leiknes. Mini-review: Novel non-destructive in situ biofilm characterization techniques in membrane systems. Desalination and Water Treatment, 57(48–49) (2016) 22894–22901.

[30] G. Massons, G. Gilabert-Oriol, V. Garcia-Molina, Method for visualizing and quantifying biofilm on solid surfaces, U.S. Patent Application 17/415,356, 2022.

[31] T. O. Peulen, K. J. Wilkinson. Diffusion of nanoparticles in a biofilm. Environmental Science & Technology, 45(8) (2011) 3367–3373.

[32] J. S. Vrouwenvelder, J. A. M. Van Paassen, L. P. Wessels, A. F. Van Dam, S. M. Bakker. The membrane fouling simulator: A practical tool for fouling prediction and control. Journal of Membrane Science, 281(1–2) (2006) 316–324.

[33] G. Massons-Gassol, G. Gilabert-Oriol, R. Garcia-Valls, V. Gomez, T. Arrowood. Development and application of an accelerated biofouling test in flat cell. Desalination and Water Treatment, 57(48–49) (2016) 23318–23325.

[34] W. S. Rasband, IMAGEJ, U.S. National Institutes of Health, Bethesda, MD, http://rsb.info.nih.gov/ij/, 1997–2006.

[35] S. Asapu, S. Pant, C. L. Gruden, I. C. Escobar. An investigation of low biofouling copper-charged membranes for desalination. Desalination, 338 (2014) 17–25.

[36] K. Danis-Wlodarczyk, T. Olszak, M. Arabski, S. Wasik, G. Majkowska-Skrobek, D. Augustyniak, G. Gula, Y. Briers, H. B. Jang, D. Vandenheuvel, K. A. Duda. Characterization of the newly isolated lytic bacteriophages KTN6 and KT28 and their efficacy against Pseudomonas aeruginosa biofilm. PloS One, 10(5) (2005) e0127603.

[37] A. Zaky, I. Escobar, A. M. C. Motlagh. Gruden Determining the influence of active cells and conditioning layer on early stage biofilm formation using cellulose acetate ultrafiltration membranes. Desalination, 286 (2012) 296–303.

[38] E. Morgenroth, K. Milferstedt. Biofilm engineering: Linking biofilm development at different length and time scales. Reviews in Environmental Science and Bio/Technology, 8(3) (2009) 203–208.

[39] M. Staal, A. Siddiqui, M. Van Loosdrecht, J. S. Vrouwenvelder. Biofouling patterns in spacer filled channels: High resolution imaging for characterization of heterogeneous biofilms. Desalination and Water Treatment, 80 (2017) 1–10.

[40] G. Massons-Gassol, G. Gilabert-Oriol, V. Gomez, R. Garcia-Valls, V. G. Molina, T. Arrowood. Method for distinguishing between abiotic organic and biological fouling of reverse osmosis elements used to treat wastewater. Desalination and Water Treatment, 83 (2017) 1–6.

Chapter 12
Why preventing biofouling matters

Energy saving in municipal drinking water application with new innovation, the FilmTec™ EXLE-440 reverse osmosis membrane element

This study highlights the advantages of a new reverse osmosis (RO) element designed for treating municipal drinking water applications, the FilmTec™ EXLE-440 RO membrane element. This new innovation is an upgrade to the FilmTec™ XLE-440 RO membrane element developed with a focus on municipal drinking water applications, and offers low energy and chemical consumption and more stable operation due to higher biofouling resistance. The study presented here includes single-element experiments with synthetic solutions and continuous operation with River Ebro water. This chapter demonstrated that the elements of the new innovation have a 10% lower energy consumption and require 30% less chemical cleanings. This enables extending the operating time by 56%, while also reducing the feed-concentrate pressure drop by 55%.

12.1 Introduction

RO is an established technology for various applications, including seawater desalination, treatment of wastewater, and production of drinking water from surface and groundwater [1]. The presence and awareness of emerging pollutants in surface and groundwater has favored the use of RO technology, as it offers a high rejection for a wide range of pollutants [2–3]. Moreover, RO membrane technology has been established as the most effective technology for desalinating water, as compared to thermal-based desalination – it consumes reliable lower energy [4].

Biological fouling, also known as biofouling, is a phenomenon where bacteria adhere to the RO membrane, and starts to build a biofilm [5]. Biofouling is caused by extracellular polymeric substances (EPS). These stick to the membrane and are a key enabler for promoting the buildup and development of a biofilm on the RO membrane element [6]. Transparent exopolymer particles (TEP) as well as protobiofilms play a key role in the development of the biofilm [7].

Acknowledgments: The author would like to acknowledge the contributors of this chapter: Claudia Niewersch, Guillem Gilabert-Oriol, and Brittany Fisher. The author would like to thank all members of the DuPont Water Solutions team for their outstanding expertise in performing these experiments.

This chapter is in the process of being submitted for publication, and was presented at the European Desalination Society (EDS) conference in Sharm El-Sheikh (Egypt) in May 2024.

https://doi.org/10.1515/9783111639079-012

Biofouling is a big challenge that affects polyamide-based membranes as they are not chlorine tolerant and so biofouling cannot be cleaned effectively using chlorine, like it is currently being done with ultrafiltration membranes [8]. So, biological fouling still remains the key problem in RO membranes systems, as it decreases the RO system's performance [9]. When an RO membrane is affected by biological fouling, the membranes need to be cleaned. Frequent chemical cleanings of the RO membranes can potentially shorten the membrane's lifetime [10].

Different solutions have been proposed to improve the RO membranes to make them more resistant to biofouling [11]. Developing biofouling-resistant feed spacers has been identified as a key component for addressing the challenge that biofouling represents [12]. At the same time, saving energy and reducing chemical consumption are also increasingly important in the water treatment sector [13]. Reducing energy consumption and reducing the number of chemical cleanings are important because they improve process stability. This new innovation, the FilmTec™ EXLE-440 membrane, has been developed, taking into account the important role that a low pressure drop feed spacer offers in addressing the challenge that biofouling presents, as well as enabling energy savings in an RO membrane system. This chapter highlights how this new innovation can reduce the impact that biofouling in RO systems has while reducing the number of chemical systems as well as showing energy savings.

12.2 Materials and methods

12.2.1 Water characterization

The energy savings and lower differential pressure were demonstrated by performing tests under standard conditions in an element configuration with synthetic solutions in recirculation mode and a 50-day trial with continuous once-through filtration of River Ebro water as feed water coming from l'Ampolla, Catalonia, Spain. This water undergoes a conventional pretreatment before reaching the Global Water Technology Center that DuPont has in Tarragona, Catalonia, Spain. The experiments with synthetic solutions were performed at 15.5 bar pressure, 2,000 ppm NaCl as feed solution, pH 8, 25 °C with feed flow in the range of 4 to 13 m³/h. The experiment was conducted by recirculating the concentrate and the permeate back into the feed tank to maintain constant conditions. The characterization of the River Ebro water with all its main ions, pH, total organic carbon (TOC), and total dissolved solids (TDS) can be seen in in Table 12.1.

Table 12.1: River Ebro water, feed water of continuous experiment.

Parameter	Average concentration	Standard deviation concentration
Li (mg/L)	0.0201	0.002602
Potassium (K) (mg/L)	3.71	0.6321
Sodium (Na) (mg/L)	126	21.37
Magnesium (Mg) (mg/L)	24.8	3.589
Calcium (Ca) (mg/L)	120	13.73
Strontium (Sr) (mg/L)	1.78	0.2735
Barium (Ba) (mg/L)	0.0305	0.00297
Iron (mg/L)	<LOQ	
Nitrate (NO3) (mg/L)	7.90	1.412
Chloride (Cl) (mg/L)	170	32.66
Sulfate (SO4) (mg/L)	235	40.91
Br (mg/L)	0.0901	0.03266
HCO_3^- (mg/L)	193	10.18
CO_3^{2-} (mg/L)	0.256	0.08934
SiO2 (mg/L)	4.01	1.370
pH	7.73	0.1542
Conductivity	1390	176.3
TDS (mg/L)	888	95.97
TOC (mg/L)	1.47	0.2597

12.2.2 Membranes used

The new innovation, the FilmTec™ EXLE-440 RO membrane element, represents an upgrade over the existing FilmTec™ XLE-440 RO membrane. This is an 8-inch spiral-wound RO membrane element. In particular, the new innovation can reduce the RO system operating pressure by 10%. This directly leads to a reduction of energy consumption by 10%. Moreover, the new innovation is able to offer a reduction of feed-concentrate pressure drop of 55%. Finally, thanks to its innovative fouling-resistant design, the new innovation can reduce the number of chemical cleanings done per year by 30%. This allows plant operators to extend the operation of the RO system by 56% without needing a chemical cleaning. The specification of the FilmTec™ XLE-440 RO membrane and the new innovation can be found in Table 12.2. It should be noted that these membranes are tested in standard test conditions. These are with recirculation, in a synthetic solution of 2,000 mg/L NaCl, 125 psi (8.6 bar), 25 °C, and 15% water recovery. Additionally, stabilization of salt rejections is typically achieved within 24–48 h of continuous use. As can be seen from this table, both elements have the same specifications on an individual element basis. Therefore, the advantages of this new innovation can be observed when the elements are used at a system level, as well as when fouling, and specifically biofouling, starts to affect the whole RO membrane system.

Table 12.2: FilmTec™ XLE-440 membrane and new the innovation, the FilmTec™ EXLE-440 membrane elements' specifications.

Membrane	Active area (ft^2)	Feed spacer (mil)	Permeate flow (gpd)	Stabilized salt rejection	Minimum salt rejection
FilmTec™ XLE-440 membrane	440	28	14,000	99.0%	97.0%
FilmTec™ EXLE-440 membrane	440	28	14,000	99.0%	97.0%

12.2.3 Experimental plant

This comparison was done at the Global Water Technology Center that DuPont Water Solutions has in Tarragona, Catalonia, Spain. For this test, five elements of the new innovation, the FilmTec™ EXLE-440 RO membrane element, were operated in parallel with six elements of the type FilmTec™ XLE-440 RO membrane. Both lines had a feed flow of 9.5 m^3/h at a permeate recovery of 48.2%. The temperature during the two months of operation was between 21 and 31 °C. A scheme of the pilot plant used for the experiment is shown in Figure 12.1.

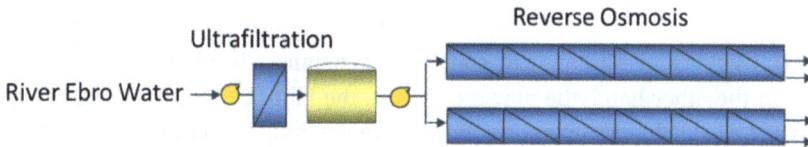

Figure 12.1: Scheme of the pilot plant used for the demonstration trial.

12.3 Results and discussion

12.3.1 Pressure drop

The results of the experiments in single-element configuration with synthetic solutions in recirculation, where the feed-concentrate pressure drop was measured as a function of the average feed-concentrate flow, are shown in Figure 12.2. Observing this plot, it can be seen that the new innovation, the FilmTec™ EXLE-440 membrane, showed a lower pressure drop in all the tested ranges.

It can also be seen from Figure 12.3 that the required cleanings due to pressure drop increase, which is caused by biofouling, can be reduced by 30%, which leads to 56% more days of operation before a cleaning is required. Moreover, it can be seen how the feed-concentrate pressure drop is reduced by 55%. This can be seen, as the

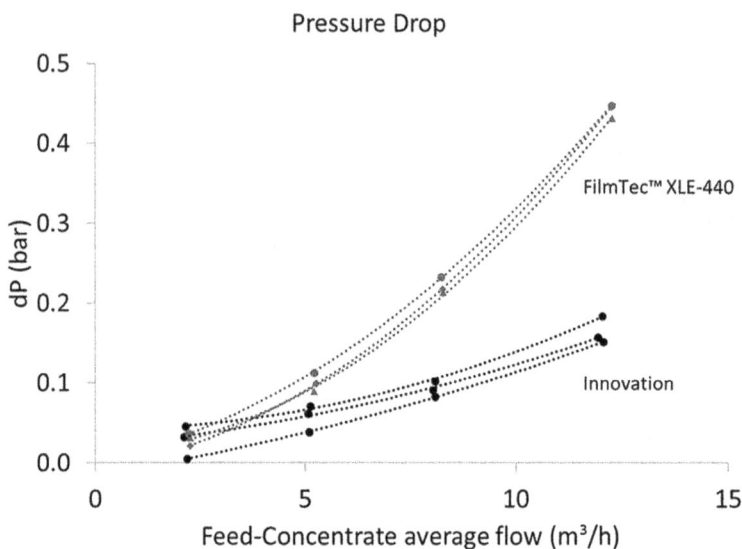

Figure 12.2: Feed-concentrate pressure drop measured for the new innovation, the FilmTec™ EXLE-440 membrane, compared with the FilmTec™ XLE-440 membrane in single elements for synthetic solutions under standard test conditions.

FilmTec™ XLE-440 RO membrane reached the third chemical cleaning after 32 days of operation. On the other hand, the new innovation, the FilmTec™ EXLE-440 RO membrane element, enabled more operating time before reaching its third chemical cleaning after 50 days of operation.

12.3.2 Feed pressure

During the continuous operation with River Ebro Water and five elements of each type, side by side, the new innovation, the FilmTec™ EXLE-440 membrane, confirmed the results of the single-element test. As can be observed during the stabilized period at the beginning of the operation, DuPont™ FilmTec™ XLE-440 RO membrane showed a feed pressure of 4.7 bar, while the new innovation showed a feed pressure of 4.2 bar. This represents a reduction on feed pressure of 10%, which corresponds to energy savings of 10%. This behavior can be seen in Figure 12.4. It is worth noting that the new innovation was able to show a reduction of 55% in the feed-concentrate pressure drop. This represents a contributing factor to energy savings. This makes the FilmTec™ XLE-440 RO membrane a low-pressure element, especially in operation with river water, and therefore, this 0.5 bar lower pressure drop results in 10% energy savings for the new innovation, the FilmTec™ EXLE-440 RO membrane element, in comparison to FilmTec™ XLE-440 RO membrane, as can be observed from the plot.

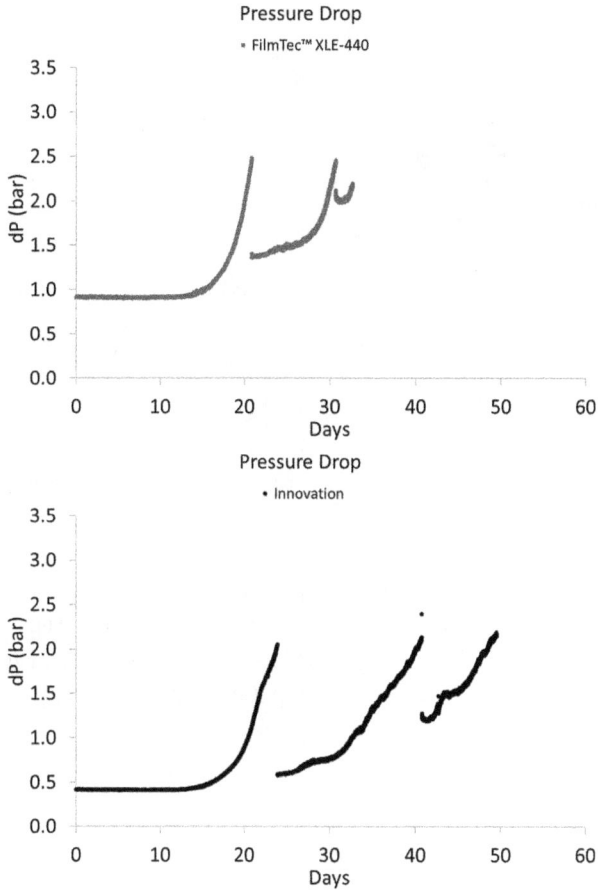

Figure 12.3: Feed-concentrate pressure drop measured during continuous operation of the new innovation, the FilmTec™ EXLE-440 membrane, in parallel with DuPont™ FilmTec™ XLE-440 elements operated with River Ebro water.

12.3.3 Salt rejection

The different salt rejections for each species while being operated with real Ebro River brackish in one through operation can be observed in Figure 12.5. There, it can be seen that the elements of the new innovation, the FilmTec™ EXLE-440 RO membrane element, are able to maintain a good rejection of the critical solutes, while at the same time, as has been shown before, providing a significant reduction in energy, in comparison to FilmTec™ XLE-440 RO membrane. On average, it can be seen that sodium rejection is maintained at 96.0%, chloride rejection is maintained at 97.5%,

Pressure

• Innovation   • FilmTec™ XLE-440

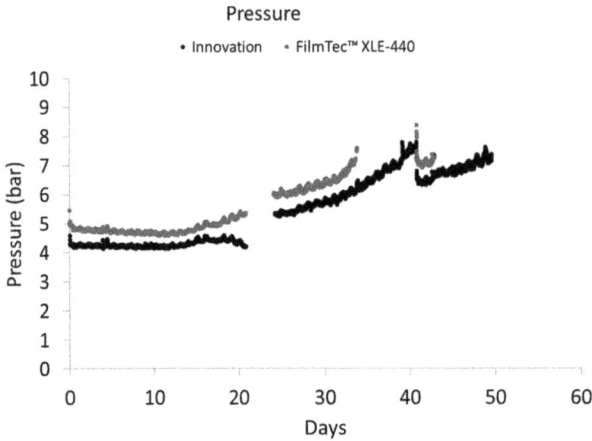

Figure 12.4: Energy savings for the new innovation, the FilmTec™ EXLE-440 membrane, demonstrated in continuous operation.

calcium rejection is maintained at 99.8%, magnesium rejection is also maintained at 99.8%, total dissolved solids are maintained at 98.4%, and total organic carbon rejection is maintained at 88.9%. These are all very good rejections for treating brackish water, and it enables maintaining a lower energy consumption while providing permeated water of high quality.

12.4 Conclusions

This study highlights the advantages of a new RO element designed for treating municipal drinking water applications, the FilmTec™ EXLE-440 RO membrane element. The new innovation was developed with a focus on municipal drinking water applications and offers low energy and chemical consumption and more stable operation due to higher biofouling resistance. The study presented here includes single-element experiments with synthetic solutions and continuous operation with River Ebro water. This chapter demonstrated that the elements of the new innovation need 10% lower energy consumption and require 30% less chemical cleanings. This results in extending the operating time without cleanings by 56%. At the same time, the operating feed-concentrate pressure drop could be reduced by 55%.

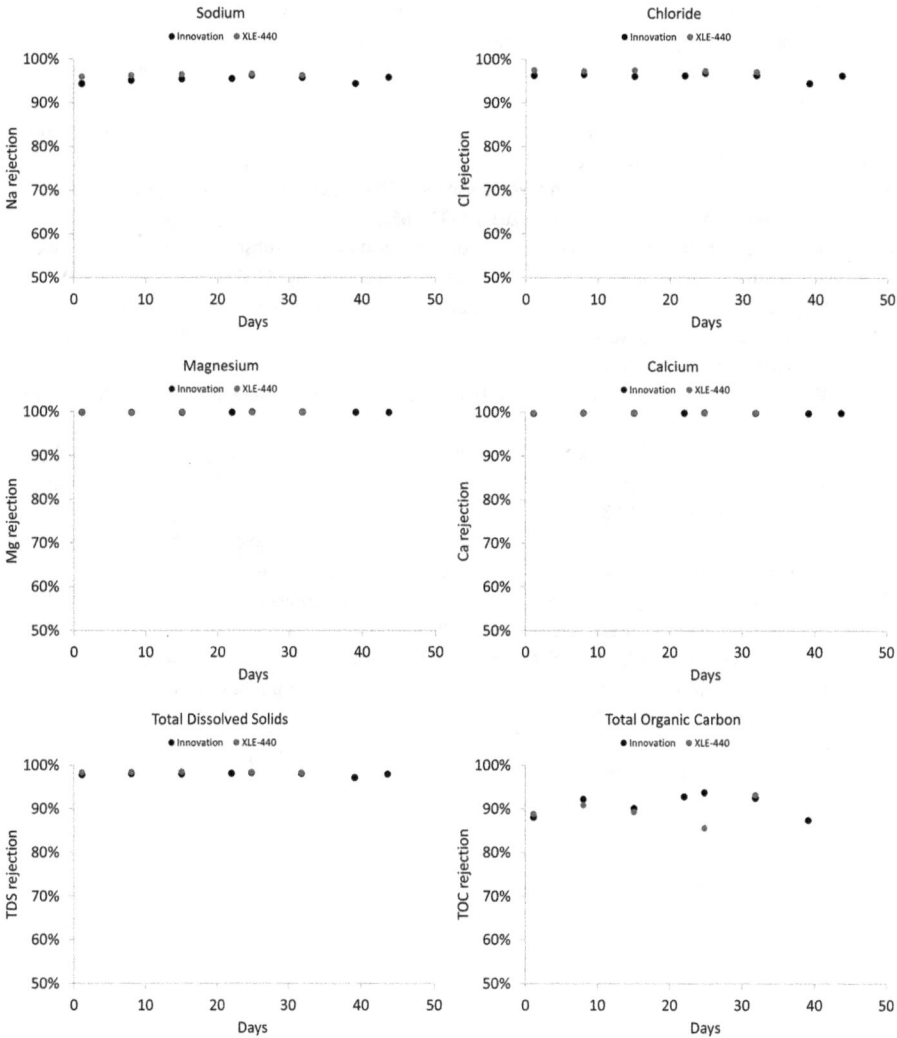

Figure 12.5: Rejection for main ions, total dissolved solids, and TOC for the FilmTec™ XLE-440 membrane, and the new innovation, the FilmTec™ EXLE-440 membrane.

References

[1]   S. Y. Pan, A. Z. Haddad, A. Kumar, S. W. Wang. Brackish water desalination using reverse osmosis and capacitive deionization at the water-energy nexus. Water Research, 183 (2020) 116064.

[2]   V. Yangali-Quintanilla, S. K. Maeng, T. Fujioka, M. Kennedy, Z. Li, G. Amy. Nanofiltration vs. reverse osmosis for the removal of emerging organic contaminants in water reuse. Desalination and Water Treatment, 34(1–3) (2011) 50–56.

[3]   N. Konradt, J. G. Kuhlen, H. P. Rohns, B. Schmitt, U. Fischer, T. Binder, V. Schumacher, C. Wagner, S. Kamphausen, U. Müller, F. Sacher. Removal of trace organic contaminants by parallel operation of reverse osmosis and granular activated carbon for drinking water treatment. Membranes, 11(1) (2021) 33.

[4]   C. Fritzmann, J. Löwenberg, T. Wintgens, T. Melin. State-of-the-art of reverse osmosis desalination. Desalination, 216(1–3) (2007) 1–76.

[5]   H. Maddah, A. Chogle. Biofouling in reverse osmosis: Phenomena, monitoring, controlling and remediation. Applied Water Science, 7 (2017) 2637–2651.

[6]   M. Herzberg, S. Kang, M. Elimelech. Role of extracellular polymeric substances (EPS) in biofouling of reverse osmosis membranes. Environmental Science & Technology, 43(12) (2009) 4393–4398.

[7]   V. Nagaraj, L. Skillman, D. Li, G. Ho. Review–Bacteria and their extracellular polymeric substances causing biofouling on seawater reverse osmosis desalination membranes. Journal of Environmental Management, 223 (2018) 586–599.

[8]   E. M. Hoek, T. M. Weigand, A. Edalat. Reverse osmosis membrane biofouling: Causes, consequences and countermeasures. Npj Clean Water, 5(1) (2022) 45.

[9]   D. Suresh, P. S. Goh, A. F. Ismail, T. W. Wong. Insights into biofouling in reverse osmosis membrane: A comprehensive review on techniques for biofouling assay. Journal of Environmental Chemical Engineering, 11(3) (2023) 110317.

[10] A. Matin, Z. Khan, S. M. J. Zaidi, M. C. Boyce. Biofouling in reverse osmosis membranes for seawater desalination: Phenomena and prevention. Desalination, 281 (2011) 1–16.

[11] S. Zhao, Z. Liao, A. Fane, J. Li, C. Tang, C. Zheng, J. Lin, L. Kong. Engineering antifouling reverse osmosis membranes: A review. Desalination, 499 (2021) 114857.

[12] J. S. Vrouwenvelder, D. G. Von Der Schulenburg, J. C. Kruithof, M. L. Johns, M. C. M. Van Loosdrecht. Biofouling of spiral-wound nanofiltration and reverse osmosis membranes: A feed spacer problem. Water Research, 43(3) (2009) 583–594.

[13] V. G. Gude. Energy consumption and recovery in reverse osmosis. Desalination and Water Treatment, 36(1–3) (2011) 239–260.

Chapter 13
Limiting nutrients to prevent biofouling in brackish water

Testing of DuPont™ B-Free™ technology in brackish water environment in Tarragona

This chapter showcases two different experiments that were able to show how with the use of DuPont™ B-Free™ pretreatment, biofouling could be fully prevented in brackish water. To prevent biological fouling, no chemicals are used, and only a simple backwash in the DuPont™ B-Free™ column is required. This chapter also shows that the DuPont™ B-Free™ column grows biomass, which is able to capture essential nutrients, and therefore prevent biofouling in the downstream reverse osmosis membrane. The correlation between this biomass thickness growth and the additional pressure drop increase caused by this biomass is also established. The biofouling prevention is confirmed through operating data, preventing up to nine chemical cleanings due to biofouling, as well as through visual inspection of the membrane surface and the feed spacer. This is also confirmed by analyzing the main parameters that are associated with fouling on the membrane and feed spacer surfaces. Additionally, it is shown that the DuPont™ B-Free™ is capable of removing more than 99.5% of the total suspended solids. So, the DuPont™ B-Free™ is shown to be a reliable and simple technology to prevent biofouling in downstream reverse osmosis systems.

13.1 Introduction

13.1.1 Biofouling

Biofouling is the most challenging type of fouling to manage in reverse osmosis (RO) systems due to the propensity of microorganisms to multiply rapidly in the non-sterile environments of commercial plants [1]. Microorganisms preferentially colonize the membrane feed spacer and, to a lesser extent, the membrane surface, generating biofilms on these surfaces. Bacterial biofilm formation is initiated by the attachment of free-floating bacteria to the surface [2]. The biofilm matrix, comprised of Extracellular Polymeric Substances (EPS), is a sticky and hydrated mixture of carbohydrates

Acknowledgments: The author would like to acknowledge the contributors of this chapter: Gerard Massons, Guillem Gilabert-Oriol, and Marc Slagt. The author would like to thank all members of the DuPont Water Solutions team for their outstanding expertise in performing these experiments.

This chapter was originally published with the following reference: DuPont™ B-Free™ Pre-treatment demonstrates its efficiency in River water, DuPont™ Water Solutions, Form No. 45-D03625-en, (2021).

https://doi.org/10.1515/9783111639079-013

and proteins, immobilizing and supporting microbial cells as they grow [3]. EPS enables microbial cell attachment to both the membrane and the feed spacer surfaces, further contributing to biofilm formation and the gradual clogging of the feed-concentrate channel. This results in the reduction of the channel's void space, leading to an increase in water resistance and an exponential pressure drop [4], thus reducing the efficiency of the reverse osmosis module and increasing energy consumption. EPS also provides considerable mechanical and chemical stability to bacteria, making biofilms challenging to clean [5].

13.1.2 Methods to mitigate biofouling

Polyamide-based membranes are sensitive to oxidizing agents such as chlorine, which significantly limits the use of chemicals for preventing bacterial growth in feed water [6]. Furthermore, conventional pretreatment methods, such as coagulation, flocculation, ultrafiltration, or cartridge filters, are inefficient in removing biofouling potential from feed water [7].

Once biofilms have accumulated on the surfaces of reverse osmosis systems, they are notoriously challenging to prevent and remove [8]. Chemical cleaners can hydrolyze the polysaccharides and proteins from the EPS matrix to disperse the fouling layers. However, the efficacy of cleaning methods for biofouling is limited, even under harsh conditions [9]. However, the harsh cleaning of biofouling is insufficient and would result in the rapid regrowth of biofilms after each chemical cleaning CIP [10].

One alternative for combating bacterial growth is metabolic inactivation, which involves in-line water treatment before it reaches the RO modules, to minimize the risk of biofilm regrowth. Physical and chemical inactivation are the two available options [11]. Biocidal agents are effective in preventing biofouling formation. However, most used biocides, such as hypochlorite and chloramines, are oxidizing compounds that are harmful to polyamide membranes over time. Only nonoxidizing biocides effectively control bacterial activity without harming thin-film composite RO membranes. However, it is important to consider the severe environmental and health risks associated with biocides when deciding their implementation.

13.1.3 Biofouling prevention technology

DuPont™ B-Free™ is a pretreatment technology that aims to alleviate the limitations of currently available methods for preventing reverse osmosis biofouling. It provides a compact, efficient, chemical-free, and robust solution to the problem. This vessel-based technology eliminates the harmful effects of biofouling, while creating an instant and sustained biostatic environment for downstream RO operations [12]. Unlike traditional methods, DuPont™ B-Free™ does not require chemicals during its opera-tion and can resist upstream upsets. The technology employs three different media, each with a specific purpose in the process, as depicted in Figure 13.1.

The idea of DuPont™ B-Free™ technology is inspired in the concept described in this book, where it is observed that biofouling mainly develops on the lead elements. Additionally, it is also observed that biofouling needs nutrients to develop, and more specifically, it needs easily assimilable nutrients. In this book, the importance of dos-ing assimilable nutrients to quickly promote biofouling has been described. This bio-fouling prevention technology uses these two concepts in a novel approach, with the idea that biofouling is growing in the wrong place. So, instead of making it grow in the reverse osmosis membrane, where it is very challenging to clean, it grows in the DuPont™ B-Free™ technology, where it is easy to control its growth and maintain it in a well-controlled place. In order to clean it, only water and air is needed. DuPont™ B-Free™ technology uses spheric media of 750 – 950 µm. Spheric media is very easy to clean as it is completely mobile, as opposed to cleaning a feed spacer that is fixed and trapped in a reverse osmosis membrane.

Therefore, DuPont™ B-Free™ technology uses three well established and synergis-tic mechanisms to create a downstream biostatic environment, so that it remains very challenging for reverse osmosis to develop biofouling.

The first mechanism happens in the Bio-stratum layer. This layer contains spheric media between 750 – 950 µm, and can be around 30 cm thick. It is designed so that bacteria can adhere in the free-void spaces between the spheric media, colonize it, and start building a biomass. This biomass acts as a biological filter that removes the majority of assimilable nutrients. Additionally, this Bio-stratum layer acts as a sieving mechanism, so that the majority of total suspended solids (TSS) can be removed, in case there is any present. These two mechanisms contribute to creating a downstream biostatic environment, so that it is more difficult for the reverse osmosis located after this pretreatment to grown biofouling.

The second mechanism happens in the Safeguard layer. This layer contains spheric media between 300 – 1,200 µm, and can be around 50 cm thick. It is designed so that any phosphate that has not been digested in the previous Bio-stratum layer can be absorbed and eliminated in this Safeguard layer. This phosphate-polishing step contributes to make the system more reliable, as it helps in making sure the down-stream water does not have the potential to build biofouling in the reverse osmosis system. This is especially important during start up, during changes in feed water composition, or after a chemical cleaning, when the biomass might not be properly established in the Bio-stratum layer. This layer also helps in filtrating and retaining any total suspended solids that might leak, or any biomass carry over that might acci-dentally go to the downstream reverse osmosis and eventually contribute to develop biofouling.

It should be noted that between the Bio-stratum layer, and the Safeguard layer, there is a distributor. This distributor is placed between both layers so that any maintenance backwash can be done effectively, and prevent any particle coming from the biomass after a backwash to migrate from the Bio-stratum layer to the Safeguard layer. Additionally, by placing this intermediate distributor, energy can be saved during the backwash and chemical cleanings. It should be noted that it is also possible to perform a chemical cleaning (CIP) using a caustic solution. During normal operation, the system is maintained through performing backwashes using only water and air, making it a sustainable, chemical-free solution against biofouling. The vessel design can be observed in Figure 13.2. The maintenance interval is influenced by the growth rate of the biomass and the linear velocity applied for filtration. Backwashing is triggered by an increase in head loss over the media bed, caused by the accumulation of biomass in the free void fraction of the media or visual observation of the stratum height, if a sight glass is available.

Finally, the third mechanism happens in the Protective Layer. This layer contains spheric media between 2.5 and 4 cm, and can be around 15 to 30 cm thick. This media is a floating inert that is located on top of the vessel, wrapping up the distributors. It is designed so that during the backwash sequence, the biomass can easily go out of the vessel, and no media is lost during the backwash flow. This effective separation and filtration of the biomass from the media helps in sustaining the operation of the DuPont™ B-Free™ technology.

Additionally, it should also be noted that a free space, called Freeboard, needs to be left between the Protective Layer and the Bio-stratum. This free space is needed so that during the backwash, the media can be properly fluidized, and the biomass can be effectively separated from the media. This Freeboard can have a space of at least the same height as the Bio-stratum layer, and ideally, an additional 30% or more height than what the Safeguard has.

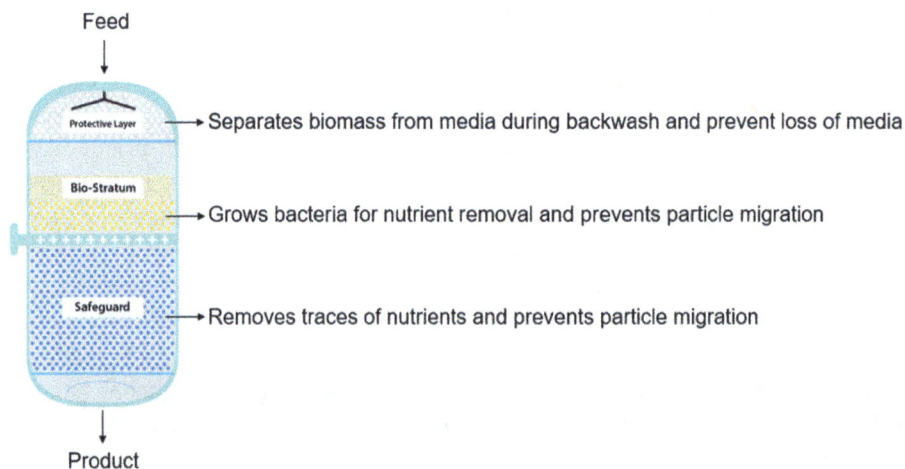

Figure 13.1: DuPont™ B-Free™ vessel layout.

The main goal of having a biofouling prevention technology such as the DuPont™ B-Free™ technology is to have a technology to prevent each type of fouling. So, the first ultrafiltration pretreatment can effectively deal with total suspended solids (TSS), and effectively remove them. This helps the downstream reverse osmosis system to operate better, with improved efficiency and reliability [13, 14]. The philosophy of the DuPont™ B-Free™ technology is similar. With this technology, biofouling can be effectively handled with this biofouling removal technology, so that the downstream reverse osmosis system can focus on doing what it is supposed to do at the highest efficiency, which is removing the total dissolved salts (TSS) from water. This combined approach, achieved with the ultrafiltration and the biofouling removal technology, contributes to maximize the efficiency and reliability of the reverse osmosis system, as the reverse osmosis stops getting fouled because of particulate fouling or biofouling, and can operate at a higher reliability. This schematic with all these three technologies incorporated is illustrated in Figure 13.2.

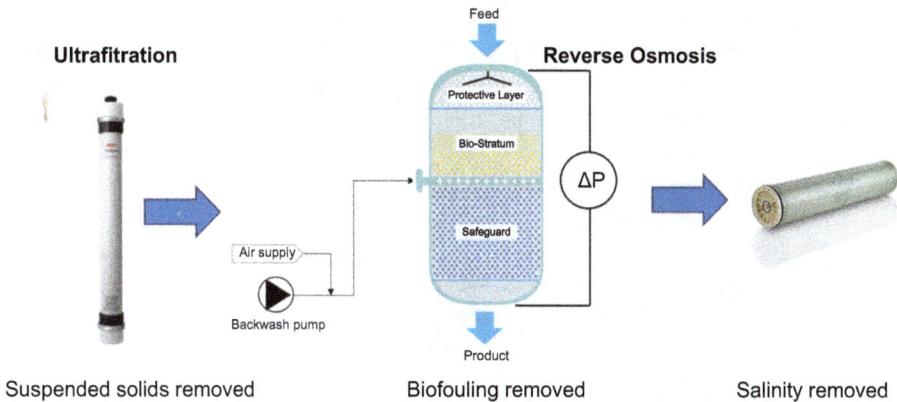

Figure 13.2: DuPont™ B-Free™ system integrated with the ultrafiltration and reverse osmosis systems.

13.2 Materials and methods

13.2.1 Brackish water characterization

River (brackish) water (total dissolved solids 1,020 mg/L and total organic carbon 1.3 mg/L) was used for the test, with the addition of nutrients, using an external dosing pump to accelerate biofilm development. Sodium acetate (0.49 mg/L), sodium nitrate (0.18 mg/L), and sodium phosphate (0.06 mg/L) were used as nutrient source in order to promote and accelerate biological fouling.

13.2.2 Pilot plant

The asset used is equipped with three plastic columns of 5 cm diameter, 19 cm^2 area, and 1 meter height. The asset is also equipped with six non-permeating flat sheet reverse osmosis membrane fouling simulators cells, with individual pressure drop measurement capabilities.

The membrane fouling simulator (MFS) units are portable and can be installed in a feed water side-stream as a stand-alone test unit or in parallel with full-scale reverse osmosis systems. A picture and a diagram of the MFS can be found in Figure 13.3. MFS are small units that simulate the feed channel of an RO element by layering the feed spacer on top of the membrane fitted in a rectangular flow cell. Water is directed through the feed spacer channel at a set flow rate, but does not permeate. The differential pressure across the MFS feed channel is monitored during the experiment. In the present study, four MFS units (MFS1, MFS2, MFS3, and MFS4) were operated in parallel. The transparent cells were assembled with a feed spacer and a membrane coupon (20 cm length and 4 cm width) in each. Manual readings from the pressure drop indicator were recorded twice per day during the course of each trial. The pressure drop as a function of time from the MFS units were compared with the respective RO element systems that ran in parallel.

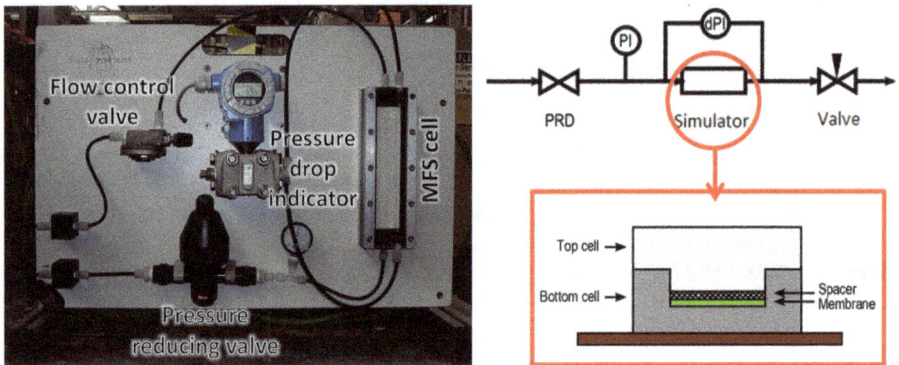

Figure 13.3: MFS simulator picture and diagram.

The columns were loaded with 1 L of media, 5 cm diameter, and 40 cm bed depth, and operated at 13 BVH and 6.6 m/h (13 L/h), as shown in Figure 13.4. The backwashes were performed at the flowrate that provided a 100% bed expansion using the feed water with no added chemicals, so that the DuPont™ B-Free™ column could be kept operating in a reliable and sustainable way.

The flat cell reverse osmosis coupons were fed either with raw water, without DuPont™ B-Free™ pretreatment used as control, or with water that is pretreated with

Figure 13.4: Layout for testing the DuPont™ B-Free™ pretreatment technology.

the DuPont™ B-Free™ technology through its column. The membranes used were FilmTec™ BW30 membranes with 28 mil standard feed spacer at a velocity of 13 L/h.

Two different experiments are performed in order to validate how effective is the DuPont™ B-Free™ technology in preventing biological fouling. The experimental setup can be observed in Figure 13.5.

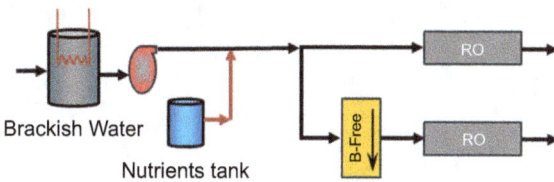

Figure 13.5: Plan schematic for testing the DuPont™ B-Free™ technology.

13.2.3 Total suspended solids removal

In order to understand how the total suspended solids (TSS) removal of DuPont™ B-Free™ technology works, the following experiment was designed to test its removal rate. Therefore, calcium carbonate ($CaCO_3$) particles were selected as a surrogate of the total suspended solids. Additionally, its particle size was also selected to be close to the typical size that bacteria exhibit, which is between 0.2 to 10 μm [15–16].

Therefore, a solution containing $CaCO_3$ microparticles with an average particle size of 7 µm, following the Gaussian curve shown in Figure 13.6 was prepared, with a total turbidity value of 473 NTU. Spheric particle size media, representative of the operating range of DuPont™ B-Free™ technology of 450 µm, 570 µm, 630 µm, and 860 µm with a uniform particle size, were tested for their particle size rejection of the total suspended solids. These particles can be visualized in Figure 13.7. The experimental setup is shown in Figure 13.8.

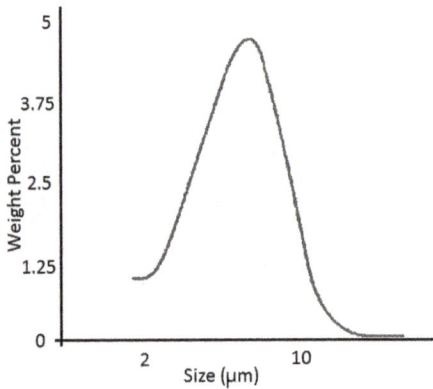

Figure 13.6: Particle size distribution of $CaCO_3$ particles.

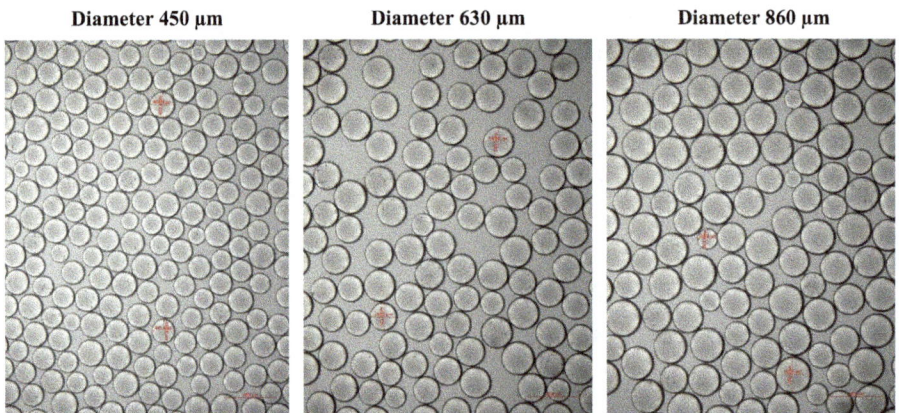

Figure 13.7: Visualization of particles with a diameter of 450 µm, 630 µm, and 860 µm.

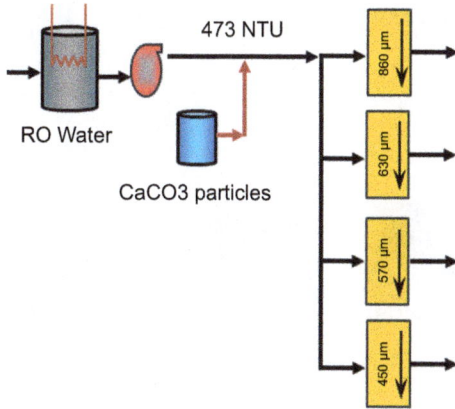

Figure 13.8: Layout for the evaluation of total suspended solids.

13.3 Results and discussion

13.3.1 First trial

13.3.1.1 Backwash

The pressure drop evolution in the DuPont™ B-Free™ column can be observed in Figure 13.9. It should be noticed that until day 24, the pressure drop in the column was not being measured. It can be seen that from day 24 to day 35, four backwashes were needed in order to effectively manage the biomass accumulation in the DuPont™ B-Free™ column. Another important point to notice is that the backwashes were efficient in restoring the pressure drop in the column to its initial value. This leads to the conclusion that the backwash is effective in managing the excess biomass and it effectively contributes to make the whole process reliable and sustainable.

13.3.1.2 Reverse osmosis

DuPont™ B-Free™ technology was able to avoid replacement of eight membranes during the 35 days that the trial lasted. These membrane replacements where done, as it was easier to replace the membrane flat sheet coupon and the feed spacer, rather than attempting to perform a chemical cleaning (CIP). This behavior is shown in Figure 13.10, where it can be seen that the reverse osmosis membrane, protected with DuPont™ B-Free™ technology, did not experienced biofouling, and its pressure drop remained stable over time. Thus, biofouling was fully prevented in the reverse osmosis system.

Pressure drop in DuPont™ B-Free™ column

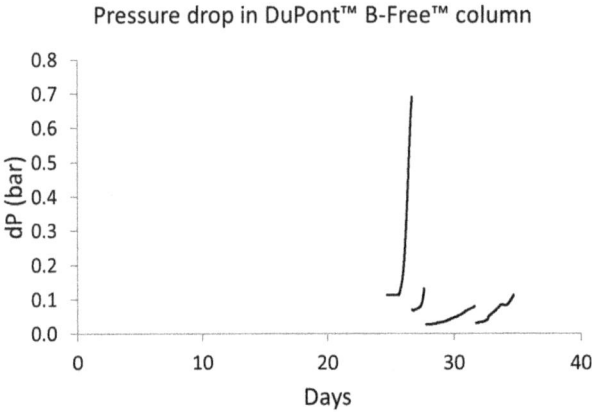

Figure 13.9: Pressure drop evolution in the DuPont™ B-Free™ column.

Pressure drop in RO without DuPont™ B-Free™ **Pressure drop in RO with DuPont™ B-Free™**

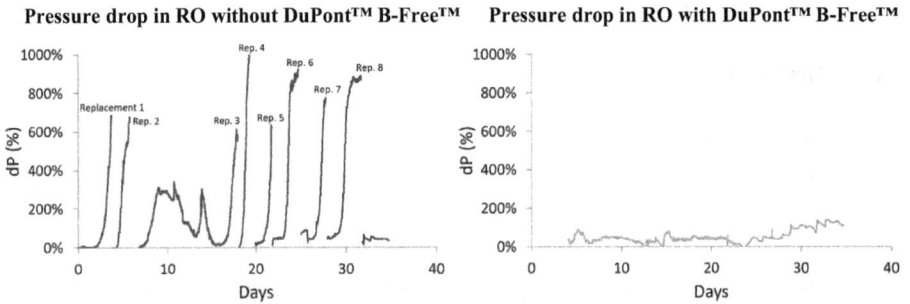

Figure 13.10: Reverse osmosis pressure drop without (left) and with (right) DuPont™ B-Free™.

13.3.1.3 Visual analysis

When the trial was over, the membrane sample, together with its feed spacer, were extracted in order to perform a visual analysis, and later a detailed fouling analysis. It should be noted that the membrane that was not protected with the DuPont™ B-Free™ pretreatment was found to have developed strong biofouling, as can be seen in Figure 13.11. This is despite the fact that this membrane operated just for 3 days, between day 32 and day 35. On the other side, the membrane sample and its feed spacer that was operated for the whole experiment with the DuPont™ B-Free™ pretreatment, did not show any noticeable trace of biofouling on its surface.

13.3.1.4 Fouling analysis

After the operation, coupons were analyzed to quantify the amount of biofouling remaining. The main parameters measured were TOC, ATP, and extrapolymeric substances (EPS), composed of proteins and carbohydrates [17]. The laboratory analysis of Figure 13.12

Figure 13.11: Visual inspection of RO coupons after autopsy.

agreed with the performance observed during the operation, which showcases that the membrane, together with its feed spacer, accumulated less biological fouling than the membrane sample that was not protected with the DuPont™ B-Free™ pretreatment.

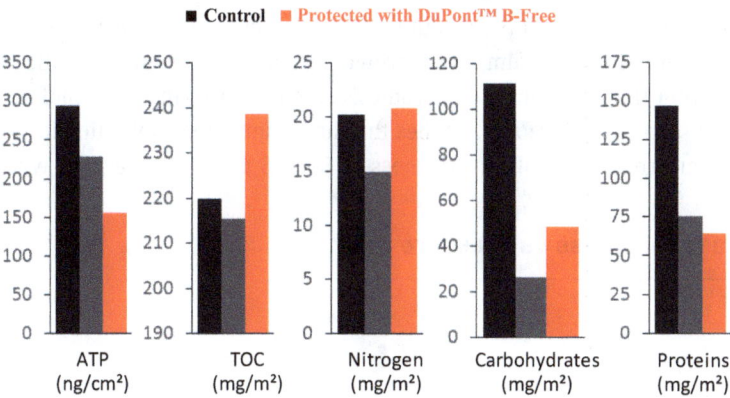

Figure 13.12: Biofouling surface concentration found in RO coupons.

13.3.1.5 Biomass growth

The biomass grew compactly and under control on the top surface of the DuPont™ B-Free™ column in the Bio-stratum layer, as shown in Figure 13.13. It should be noticed that when the thickness of the biological stratum increased, pressure drop also increased. This is attributed to the reduction in void space between the media, as it difficult for the water to flow effectively across the Bio-stratum layer. The biofilm can be observed in the picture as a white dense layer. It reached more than 2.1 cm in depth.

The correlation between the biomass thickness growth and the pressure drop increase was studied in Figure 13.14. When looking at the plot, it can be observed that biomass grows at a more or less linear rate at the same time that the pressure drop

| **dP** 0.11 bar | **dP** 0.25 bar | **dP** 0.40 bar | **dP** 0.75 bar |
| **Thickness** 0 cm | **Thickness** 0.7 cm | **Thickness** 1.4 cm | **Thickness** 2.1 cm |

Figure 13.13: Biomass growth evolution in the Bio-stratum layer of the DuPont™ B-Free™ media.

increases over time. This relationship was statistically studied and was proven statistically significant through an ANOVA analysis, with a p-value of 0.0061, which is smaller than the typically accepted confidence interval of 0.05.

This analysis is interesting, as it points out that the biomass that grows in the DuPont™ B-Free™ system is not a biofilm, but a growing biomass. It should be noticed that the typical biofilm thickness on reverse osmosis samples suffering from biofouling is around 97–114 μm [18]. Moreover, studies that have simulated biofouling in reverse osmosis system report a biofilm thickness of 80–175 μm, which agrees with

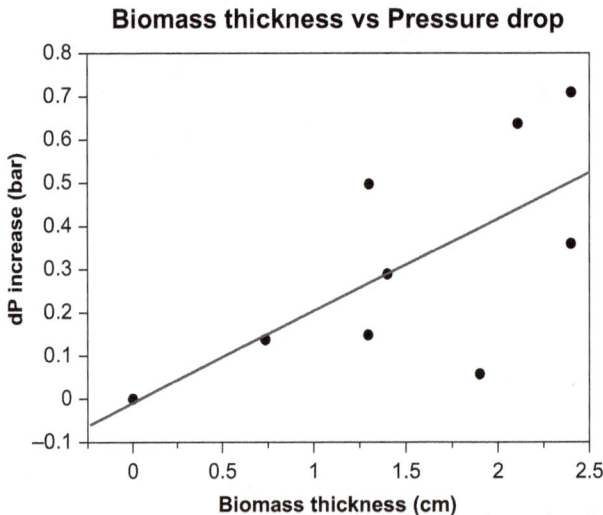

Figure 13.14: Relationship between the biomass thickness and the pressure drop increase.

these studies [19]. As can be observed, these values are quite different in order of magnitude compared to the biomass thickness of 2.1 cm obtained in this study.

A closer look of the Bio-stratum layer and the interaction of the biofilm and Du-Pont™ B-Free™ media can be observed in the microscopic picture in Figure 13.15, where biomass shows a lighter pale orange color, compared to the darker color that the DuPont™ B-Free™ media presents. It should be noted that the biomass has also been shown to be easy to separate. This is why backwash with just water and air is sufficient to keep the DuPont™ B-Free™ operating under a sustainable and reliable way, thus not requiring chemicals to operate.

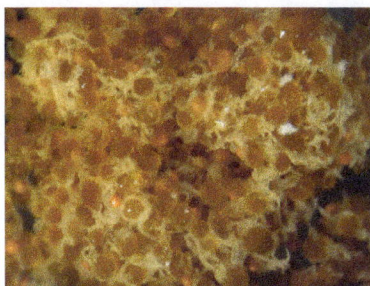

Figure 13.15: Biomass growth in the Bio-stratum layer of the DuPont™ B-Free™ media.

13.3.2 Second trial

13.3.2.1 Backwash

In this second trial, a total number of 15 backwashes were done in the 34 days that the trial lasted. This corresponds to, on average, one backwash needed to be performed every two days, when the pressure drop reached the threshold level of around 0.8 bar/m. The pressure drop in each of the two DuPont™ B-Free™ columns that where put in series could be effectively restored to the initial pressure drop with the backwash pressure, despite the aggressive increase rate observed during filtration. This behavior can be observed in Figure 13.16. It should be noticed that the backwashes were efficient in restoring the pressure drop in the column to its initial value. This leads to the conclusion that the backwash is effective in managing the excess biomass and it effectively contributes to make the whole process reliable and sustainable.

13.3.2.2 Reverse osmosis

DuPont™ B-Free™ technology was able to prevent seven membrane replacements during the 34 days that the trial lasted. These membrane replacements where done, as it was easier to replace the membrane flat sheet coupon and feed spacer, rather than attempting to perform a chemical cleaning (CIP). This behavior can be seen in Figure 13.17, where it can be seen that the reverse osmosis membrane protected with DuPont™ B-Free™ technology

Pressure drop in DuPont™ B-Free™ column

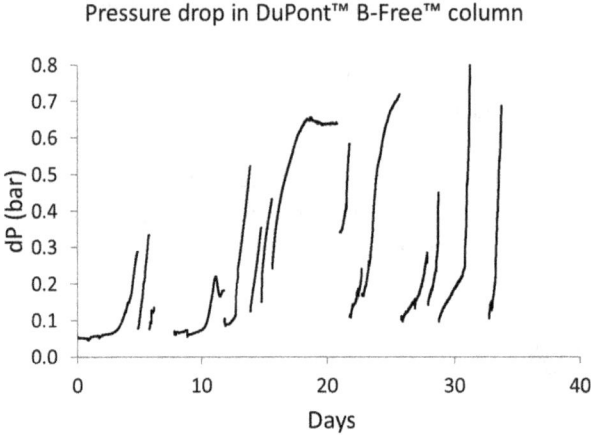

Figure 13.16: Evolution of pressure drop over the column after each backwash cycle.

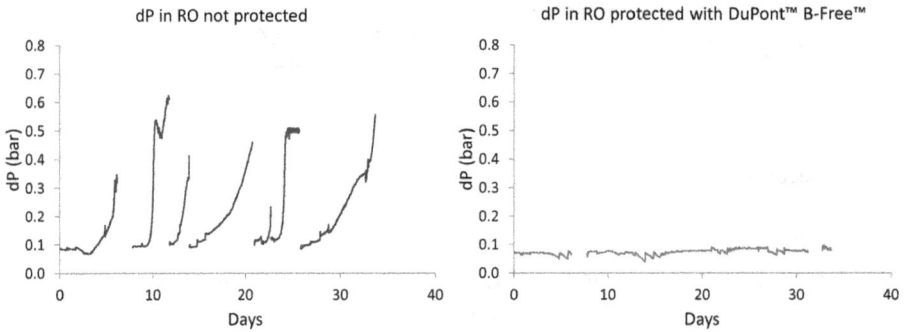

Figure 13.17: Reverse osmosis pressure drop without (left) and with (right) DuPont™ B-Free™.

did not experience biofouling, and its pressure drop remained stable over time. Thus, biofouling was fully prevented in the reverse osmosis system.

13.3.2.3 Visual analysis

When the trial was over, the membrane sample, together with its feed spacer, were extracted in order to perform a visual analysis, and later a detailed fouling analysis. It should be noted that the membrane that was not protected with the DuPont™ B-Free™ pretreatment was found to have developed strong biofouling, as can be seen in Figure 13.18. This is despite the fact that this membrane operated just for eight days, between day 26 and day 34. On the other side, the membrane sample and its feed spacer that was operated for the whole experiment with the DuPont™ B-Free™ pretreatment, did not show any noticeable trace of biofouling on its surface.

Without DuPont™ B-Free™ pretreatment

With DuPont™ B-Free™ pretreatment

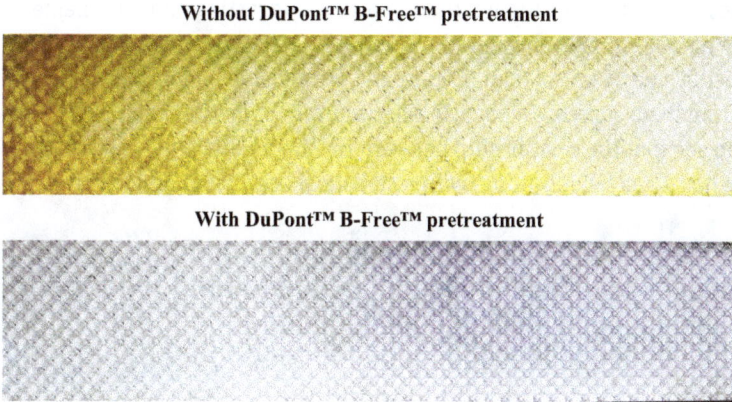

Figure 13.18: Visual inspection of reverse osmosis coupons after autopsy.

13.3.2.4 Fouling analysis

After the operation, coupons were analyzed to quantify the amount of biofouling remaining. The main parameters measured were TOC, ATP, and extrapolymeric substances (EPS), composed of proteins and carbohydrates [17]. The laboratory analysis
of Figure 13.19 agreed with the performance observed during the operation, which showcases that the membrane, together with its feed spacer, accumulated less biological fouling than the membrane sample not protected with the DuPont™ B-Free™ pretreatment.

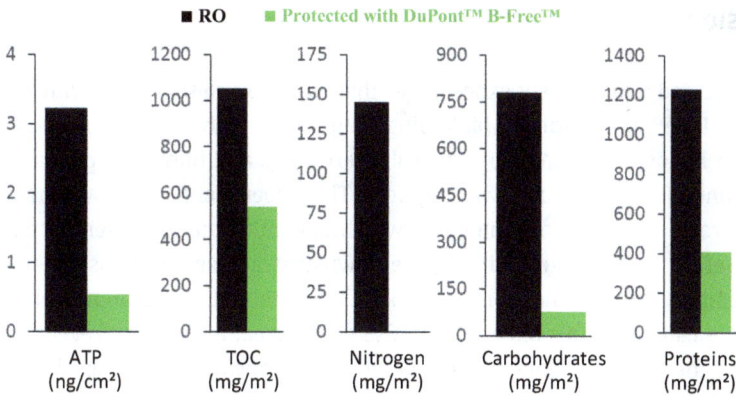

Figure 13.19: Biofouling surface concentration found in reverse osmosis membranes.

13.3.3 Media size effect on biofouling prevention

The results achieved for particle removal using a synthetic solution of calcium carbonate microparticles are shown in Figure 13.20. It can be observed how the efficiency

rapidly decreases at media size higher than 650 µm, suggesting that the bacteria might eventually be able to pass through the column bed downstream to the RO elements. With a media size of lower than 650 µm and with just 40 cm of media, almost all particles are removed, showing a particle rejection rate of greater than 99.5%. It is also interesting to see that the relationship between the media diameter and the particle rejection rate has an excellent fitting, with a statistical coefficient of determination (r^2) of 0.973.

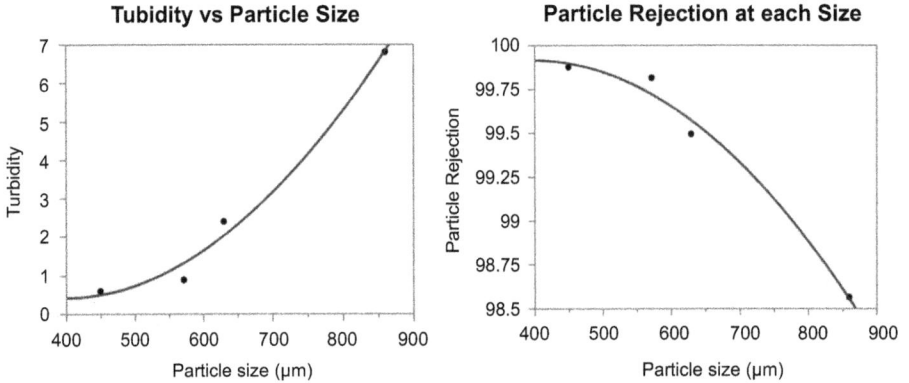

Figure 13.20: Particle filtration as a function of particle size.

13.4 Conclusions

This chapter showcased two different experiments that were able to show how with the use of DuPont™ B-Free™ pretreatment, biofouling could be fully prevented in brackish water. To prevent biological fouling, no chemicals were used, and only a simple backwash in the DuPont™ B-Free™ column was required. This experiment also showed that the DuPont™ B-Free™ column grows biomass, which is able to capture essential nutrients, and therefore prevent biofouling in the downstream reverse osmosis membrane. The correlation between the biomass thickness growth and the additional pressure drop increase caused by this biomass was also established. Biofouling prevention was confirmed by the operating data, preventing up to nine chemical cleanings due to biofouling, as well as through visual inspection of the membrane surface and the feed spacer. This was later confirmed by analyzing the main parameters that are associated with fouling on the membrane and on the feed spacer surfaces. Additionally, it was shown that the DuPont™ B-Free™ is capable of removing more than 99.5% of the total suspended solids. So, the DuPont™ B-Free™ is shown to be a reliable and simple technology to prevent biofouling in downstream reverse osmosis systems.

References

[1]   L. Javier, N. M. Farhat, P. Desmond, R. V. Linares, S. Bucs, J. C. Kruithof, J. S. Vrouwenvelder. Biofouling control by phosphorus limitation strongly depends on the assimilable organic carbon concentration. Water Research, 183 (2020) 116051.

[2]   T. Bhattacharjee, S. S. Datta. Confinement and activity regulate bacterial motion in porous media. Soft Matter, 15 (2019) 9920–9930.

[3]   M. Schulz, J. Winter, H. Wray, B. Barbeau, P. Bérubé. Biologically active ion exchange (BIEX) for NOM removal and membrane fouling prevention. Water Science and Technology: Water Supply, 17 (2017) 1178–1184.

[4]   H. Yang, B. Minuth, D. Grant Allen. Effects of nitrogen and oxygen on biofilter performance. Journal of the Air & Waste Management Association, 52 (2002) 279–286.

[5]   H. C. Flemming. Biofouling in water systems – Cases, causes and countermeasures. Applied Microbiology and Biotechnology, 76(6) (2007) 1255–1267.

[6]   L. Vanysacker, R. Bernshtein, I. F. J. Vankelecom. Effect of chemical cleaning and membrane aging on membrane biofouling using model organisms with increasing complexity. Journal of Membrane Science, 457 (2014) 19–28.

[7]   Z. Liu, K. M. Lompe, M. Mohseni, P. R. Bérubé, S. Sauvé, B. Barbeau. Biological ion exchange as an alternative to biological activated carbon for drinking water treatment. Water Research, 168 (2020) 115148.

[8]   S. West, H. Horn, W. A. M. Hijnen, C. Castillo, M. Wagner. Confocal laser scanning microscopy as a tool to validate the efficiency of membrane cleaning procedures to remove biofilms. Separation and Purification Technology, 122 (2014) 402–411.

[9]   H. Wray, M. Schulz, J. Winter, P. R. Bérubé, B. Barbeau, W. Bayless, Biological Ion Exchange for NOM Removal – Unforeseen Synergies, American Water Works Association – Water Quality Technology Conference (WQTC), (2016) 13–17

[10]   M. Clements, J. Haarhoff. Practical experiences with granular activated carbon (GAC) at the Rietvlei Water Treatment Plant. Water SA, 30 (2004) 89–95.

[11]   T. Nguyen, F. A. Roddick, L. Fan. Biofouling of water treatment membranes: A review of the underlying causes, monitoring techniques and control measures. Membranes, 2 (2012) 804–840.

[12]   B. D. Martin, L. De Kock, M. Gallot, E. Guery, S. Stanowski, J. MacAdam, B. Jefferson. Quantifying the performance of a hybrid anion exchanger/adsorbent for phosphorus removal using mass spectrometry coupled with batch kinetic trials. Environmental Technology, 39 (2018) 2304–2314.

[13]   O. Lorain, B. Hersant, F. Persin, A. Grasmick, N. Brunard, J. M. Espenan. Ultrafiltration membrane pre-treatment benefits for reverse osmosis process in seawater desalting. Quantification in terms of capital investment cost and operating cost reduction. Desalination, 203(1–3) (2007) 277–285.

[14]   F. Fatima, S. Fatima, H. Du, R. R. Kommalapati. An evaluation of microfiltration and ultrafiltration pretreatment on the performance of reverse osmosis for recycling poultry slaughterhouse wastewater. Separations, 11(4) (2024) 115.

[15]   A. I. Radu, J. S. Vrouwenvelder, M. C. M. Van Loosdrecht, C. Picioreanu. Effect of flow velocity, substrate concentration and hydraulic cleaning on biofouling of reverse osmosis feed channels. Chemical Engineering Journal, 188 (2012) 30–39.

[16]   C. Dreszer, J. S. Vrouwenvelder, A. H. Paulitsch-Fuchs, A. Zwijnenburg, J. C. Kruithof, H. C. Flemming. Hydraulic resistance of biofilms. Journal of Membrane Science, 429 (2013) 436–447.

[17]   M. Jafari, A. D'haese, J. Zlopasa, E. R. Cornelissen, J. S. Vrouwenvelder, K. Verbeken, C. Picioreanu. A comparison between chemical cleaning efficiency in lab-scale and full-scale reverse osmosis membranes: Role of extracellular polymeric substances (EPS. Journal of Membrane Science, 609 (2020) 118189.

[18]   A. Katz, A. Alimova, M. Xu, E. Rudolph, M. K. Shah, H. E. Savage, R. B. Rosen, S. A. McCormick, R. R. Alfano. Bacteria size determination by elastic light scattering. IEEE Journal of Selected Topics in Quantum Electronics, 9(2) (2003) 277–287.

[19]   W. F. Marshall, K. D. Young, M. Swaffer, E. Wood, P. Nurse, A. Kimura, J. Frankel, J. Wallingford, V. Walbot, X. Qu, A. H. Roeder. What determines cell size? BMC Biology, 10 (2012) 1–22.

Chapter 14
Limiting nutrients to prevent biofouling in wastewater

Testing of DuPont™ B-Free™ technology in a wastewater environment in Tarragona

A recent technology named B-Free™ was piloted to demonstrate its ability to prevent bio-fouling in reverse osmosis (RO) filtration. Wastewater with high organic and biological fouling potential was used as feed to an RO pilot plant using 1812 membrane elements, equipped with a DuPont™ B-Free™ column as pretreatment. The chapter aims to deter-mine and quantify the benefits achieved when applying DuPont™ B-Free™ in a challeng-ing application like municipal wastewater reclamation. DuPont™ B-Free™ showed effec-tive biofilm management under high organic load conditions, with an observable biofilm layer in the bio-stratum layer that was easily breakable with air scour and backwash. The cartridge filter operated without B-Free™ suffered a higher pressure drop increase rate, leading to more frequent replacements. The RO elements operated with DuPont™ B-Free™ showed a sustained reduction in pressure drop in all the trials. Additionally, this biofouling prevention technology helped remove organic substances such as humic sub-stances, nitrate, biopolymers, and hydrophobic carbon from the feed water. This can po-tentially decrease the organic fouling that an RO system might experience.

14.1 Introduction

14.1.1 Biofouling

Biofouling is a major issue in reverse osmosis (RO) systems, primarily due to the rapid growth of microorganisms in the non-sterile conditions common to commercial plants [1]. These microorganisms tend to accumulate on the membrane feed spacer and, to a lesser extent, on the membrane surface, where they form biofilms. The pro-cess of biofilm formation begins when free-floating bacteria attach to the surface [2]. The biofilm matrix, composed of extracellular polymeric substances (EPS), is a viscous and hydrated blend of carbohydrates and proteins that captures and supports microbial

Acknowledgments: The author would like to acknowledge the contributors to this chapter: Gerard Mas-sons, Guillem Gilabert-Oriol, and Marc Slagt. The author would like to thank the entire DuPont Water Sol-utions team for their outstanding expertise in performing these experiments.

This chapter was originally published with the following reference: DuPont™, DuPont™ B-Free™ Pre-treatment demonstrates its efficiency in municipal wastewater, DuPont™ Water Solutions, Form No. 45-D03626-en (2021).

https://doi.org/10.1515/9783111639079-014

cells as they proliferate [3]. EPS plays a crucial role in the attachment of microbial cells to both the membrane and feed spacer surfaces, promoting further biofilm development and leading to the gradual obstruction of the feed-concentrate channel. This narrowing of the channel reduces the available space for water flow, which increases resistance and results in a significant pressure drop [4]. Consequently, the performance of the RO module is compromised, and energy consumption rises. Additionally, EPS provides the biofilm with enhanced mechanical and chemical resilience, making it difficult to effectively remove during cleaning procedures [5].

14.1.2 Methods to mitigate biofouling

Polyamide-based membranes are highly susceptible to oxidative agents like chlorine, which severely restricts the use of certain chemicals for preventing bacterial growth in feed water [6]. In addition, traditional pretreatment techniques, such as coagulation, flocculation, ultrafiltration, or cartridge filtration, fail to effectively remove the biofouling potential in feed water [7].

Once biofilms settle on the surfaces of RO systems, they become particularly difficult to control and eradicate [8]. Chemical cleaning agents can break down the polysaccharides and proteins in the EPS matrix, helping to disrupt fouling layers. However, the effectiveness of these cleaning methods is limited, especially under extreme conditions [9]. In addition, harsh cleaning procedures are often inadequate and lead to rapid biofilm regrowth after each chemical cleaning cycle (CIP) [10].

To tackle bacterial growth, an alternative method is metabolic inactivation, which involves treating the water before it enters the RO modules to reduce the chances of biofilm regrowth. The two main inactivation approaches are physical and chemical methods [11]. While biocidal agents can prevent biofouling, many common biocides, such as hypochlorite and chloramines, are oxidizing substances that can damage polyamide membranes over time. Non-oxidizing biocides, however, can control bacterial activity without harming thin-film composite membranes. Still, it is essential to weigh the environmental and health hazards associated with biocides when considering their use.

14.1.3 Biofouling prevention technology

DuPont™ B-Free™ is a pretreatment technology designed to overcome the challenges of RO biofouling, offering a compact, efficient, and chemical-free solution [12]. This vessel-based technology creates a sustained biostatic environment for downstream RO operations, mitigating biofouling's harmful effects and resisting upstream upsets. It operates without chemicals, employing three synergistic mechanisms within distinct media layers to promote controlled biomass growth and nutrient removal, as depicted in Figure 14.1.

The principles behind DuPont™ B-Free™ are rooted in observations from earlier chapters of this book, where biofouling was shown to predominantly develop on lead RO elements and rely on easily assimilable nutrients. The previous chapter specifically highlights the significance of nutrient dosing in biofouling processes, laying the groundwork for this novel approach. Rather than allowing biofouling to form on reverse osmosis membranes, this technology encourages growth within the pretreatment vessel, where it is more easily managed and cleaned using water and air.

The Bio-stratum layer (750–950 µm, 30 cm thick) facilitates bacterial adhesion, biomass formation, and nutrient removal, creating a biostatic environment while sieving total suspended solids (TSS). The Safeguard layer (300–1,200 µm, ~50 cm thick) absorbs residual phosphate and filters biomass carryover, ensuring reliability during variations in feedwater or startup conditions. These layers are separated by a distributor, enhancing backwash efficiency and conserving energy. Maintenance primarily involves water and air backwashes, with chemical cleanings (CIP) as needed.

The Protective Layer (2.5–4 cm, 15–30 cm thick) ensures no media loss during backwashes, with a freeboard zone providing fluidization space for efficient biomass removal. Together, these mechanisms ensure sustainable and robust biofouling prevention for downstream RO systems.

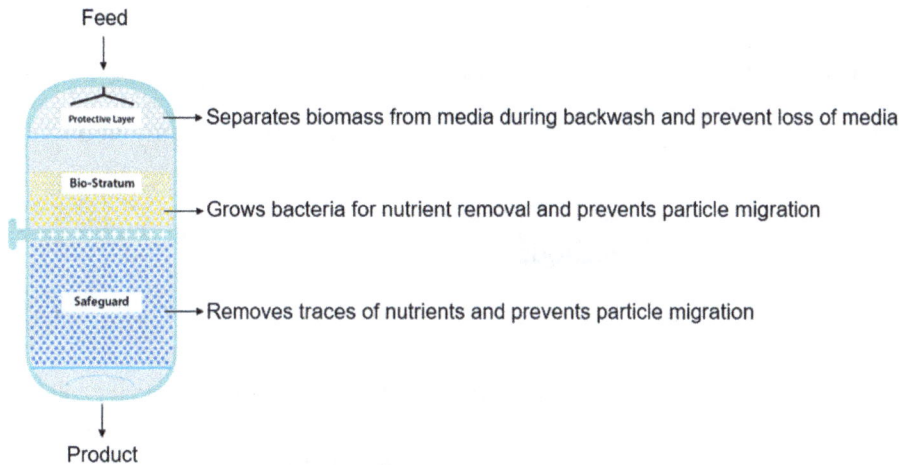

Figure 14.1: DuPont™ B-Free™ vessel layout.

The primary goal of biofouling prevention technologies, such as DuPont™ B-Free™, is to effectively mitigate various types of fouling. Ultrafiltration pretreatment plays a crucial role in eliminating TSS, which significantly enhances the efficiency and reliability of downstream RO systems [13, 14]. Similarly, DuPont™ B-Free™ technology targets biofouling, ensuring that RO systems can focus on their primary function, which is removing total dissolved salts from water while operating at peak performance. The integration of ultrafiltration, biofouling prevention, and RO represents a holistic approach to maximizing operational efficiency. By addressing particulate fouling and biofouling, this synergy ensures consistent and reliable RO performance. This interaction among technologies is illustrated in Figure 14.2.

Building on the discussions in the previous chapters, which examined the limitations of conventional pretreatment methods in mitigating biofouling in the downstream RO system, this chapter demonstrates how innovative technologies like DuPont™ B-Free™ overcome these challenges to optimize water treatment processes and effectively prevent biological fouling.

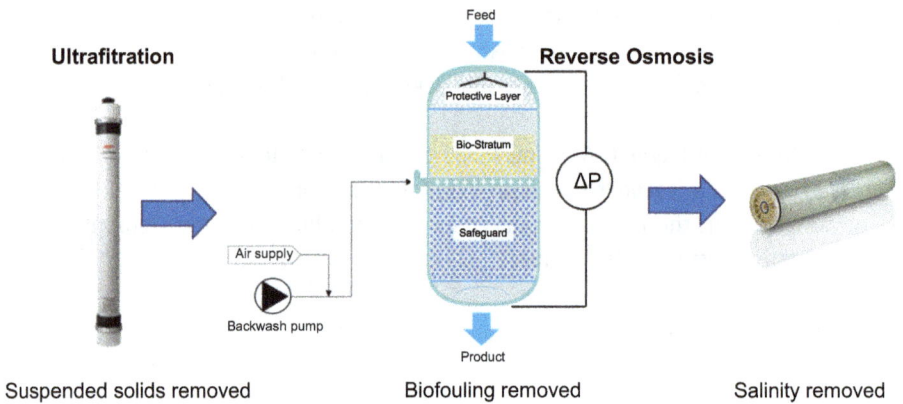

Figure 14.2: DuPont™ B-Free™ system integrated between the ultrafiltration and reverse osmosis systems.

14.2 Materials and methods

14.2.1 Wastewater characterization

Vila-seca wastewater treatment plant treats municipal wastewater from the villages of Vila-seca, La Pineda, and Salou. The intake of the water is from the effluent of the clarifier, which is then pretreated using two DuPont™ Ultrafiltration SFP-2880 XP modules. The average ionic and organic characteristics of the ultrafiltered water used for the RO experiments are summarized in Table 14.1.

Table 14.1: Wastewater feed characterization.

Parameter	Average concentration
ATP	0.03 ± 0.02 ng/L
TOC	8 ± 1.5 mg/L
Calcium	135 ± 16 mg/L
Chloride	509 ± 48 mg/L
Nitrate	63 ± 26 mg/L
Sodium	308 ± 30 mg/L
Sulfate	290 ± 49 mg/L
Conductivity	$2{,}408 \pm 442$ µS/cm
pH	7.3 ± 0.2

14.2.2 Experimental plant

The tests were performed using ultrafiltrate municipal wastewater as feed water. The biofouling reduction achieved was validated with the pilot plant described in Figure 14.3. Ultrafiltrate water is distributed into two parallel pilot units, one equipped with DuPont™ B-Free™.

The DuPont™ B-Free™ pilot used was equipped with two columns of 0.15 m internal diameter and operated in series. Each column was filled with 0.50 m of media. This resulted in 9 L of media in each one of the two different columns. The first column contained the bio-stratum layer media. The second column contained the Safeguard layer. The columns operated at a flow rate of 300 L/h and a velocity of 33 BV/h, which corresponded to a velocity of 17 m/h.

The pretreated water with DuPont™ B-Free™ is fed to up to four single-element pressure vessels, using an 1812 membrane element configuration. Each RO membrane receives 100 L/h of feed water and produced a permeate of 5 L/h. The biofouling potential of the RO elements equipped with DuPont™ B-Free, represented in the figure with red color, is compared to the RO elements fed directly with ultrafiltrate water, represented in the figure with blue color. The CF for both lines is also compared to determine the potential reduction in replacement rates due to pretreatment.

The BW trigger was the absolute pressure drop (dP) value. When the column reached around 0.6 bar/m, the BW sequence was performed. The BW process in this pilot plant is a manual process, and the conditions used included predraining of the water, a 5-min air scour at 15 m³/h, and a BW step of 60 min at 17 m/h velocity.

The two transparent columns containing the DuPont™ B-Free™ media, which are operating in series, can be seen in Figure 14.4.

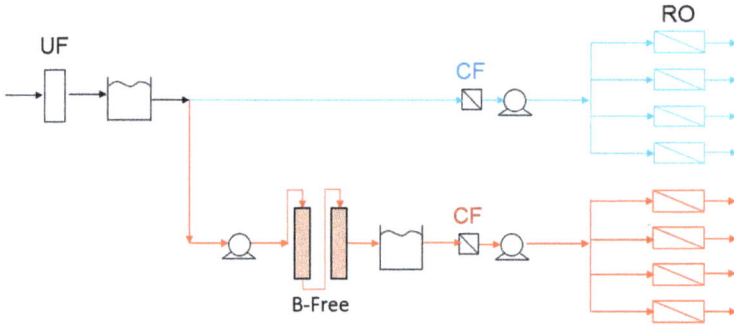

Figure 14.3: Pilot plant used for biofouling evaluation.

Figure 14.4: First (left) and second (right) columns in series with DuPont™ B-Free™ media.

14.3 Results and discussion

14.3.1 Backwashes

On average, a BW needed to be performed every 5 days, when the pressure drop reached around the 0.8 bar/m threshold level. The pressure drop in each of the two DuPont™ B-Free™ columns that were put in series could be effectively restored to the initial pressure drop with the BW pressure despite the aggressive increase rate observed during filtration. This behavior can be observed in Figure 14.5.

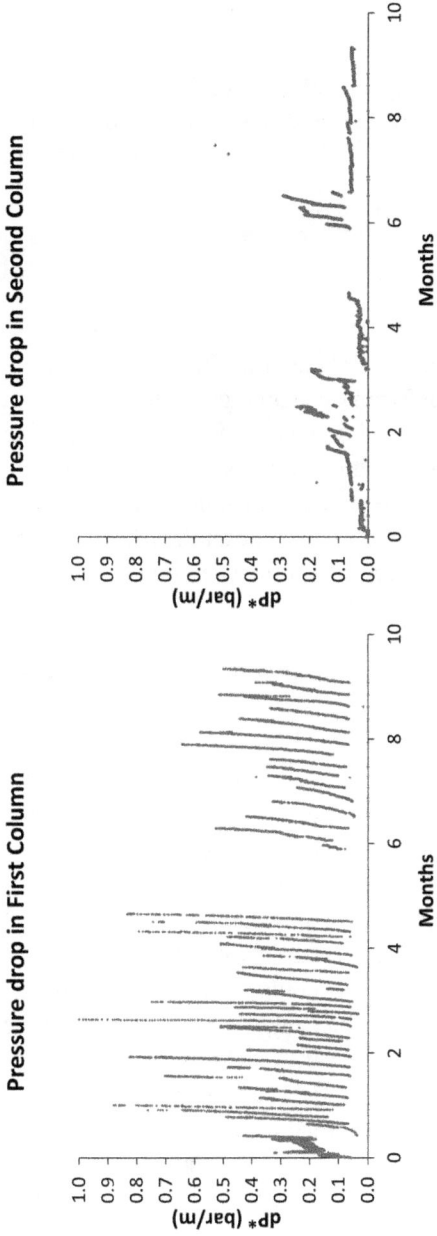

Figure 14.5: Evolution of pressure drop over the column after each backwash cycle.

14.3.2 Cartridge filter

It can be observed that the CF operated without DuPont™ B-Free™ suffered a higher pressure drop increase rate than the one protected with the biofouling prevention technology. This led to more frequent replacements of the CFs. These test results are summarized in Figure 14.6.

14.3.3 Reverse osmosis

Biological fouling is typically associated with an increase in the pressure drop of reverse osmosis elements. The reduction in biofouling potential of the feed water can be validated by monitoring the pressure drop trend of the RO elements operated with DuPont™ B-Free™ pretreated water compared to the reverse osmosis operated without DuPont™ B-Free™. Figure 14.7 shows the sustained reduction in pressure drop build-up for the elements operated with DuPont™ B-Free™ for all the trials. The efficiency is different from test to test due to adjustments in the operational conditions and seasonal effects.

14.3.4 Biofilm management

The operation with municipal wastewater allowed for the observation and validation of effective biofilm management under challenging conditions. The high nutrient content in the water caused rapid biofilm growth and triggered frequent BWs. However, as can be observed, a thick biofilm formed in the Bio-stratum layer but could be easily broken up using air scour and BW. During normal operation, the system is maintained by BW, using water and air only, making it a sustainable, chemical-free solution against biofouling. A picture of how the media looks during operation and during its maintenance BW can be seen in Figure 14.8. The maintenance interval is influenced by the growth rate of the biomass and the applied filtration linear velocity. The BW is triggered by the increase in pressure drop over the media bed caused by the accumulation of biomass in the free void fraction of the media or visual observation of the stratum height if a sight-glass is available. Air is used to break the biomass into smaller pieces, which are then easier to backwash. The biomass that grows tends to form an omelet-like disc, which can occupy up to the entire available diameter of the vessel. It should be noted that the backwash triggers the fluidization of the media, and biomass easily detaches from the media, as they are designed to have different densities.

The transparent column used also allowed visualization of the effect of the protective layer during the backwash process, enabling the removal of biomass from the column while preventing any bio-stratum layer media from being lost in the process. Figure 14.9 shows a time lapse of the process with details on the top distribution area.

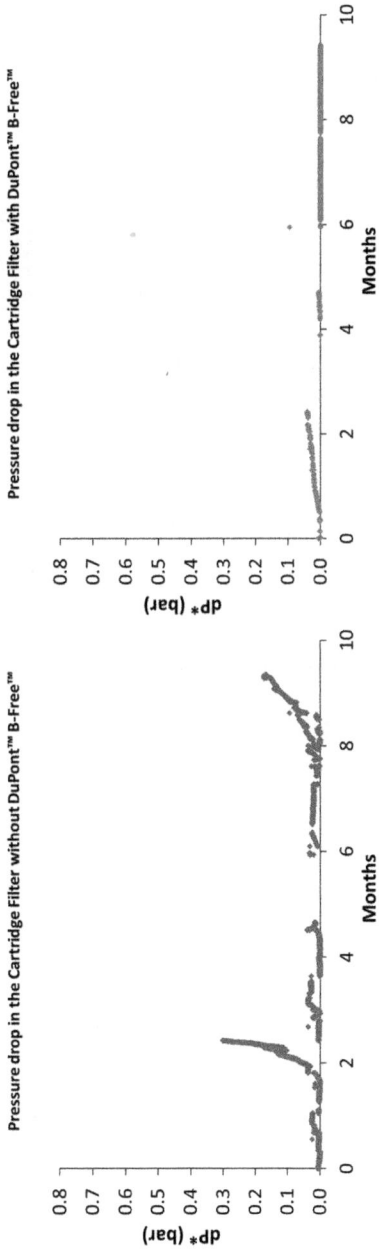

Figure 14.6: Cartridge filter pressure drop with and without DuPont™ B-Free™.

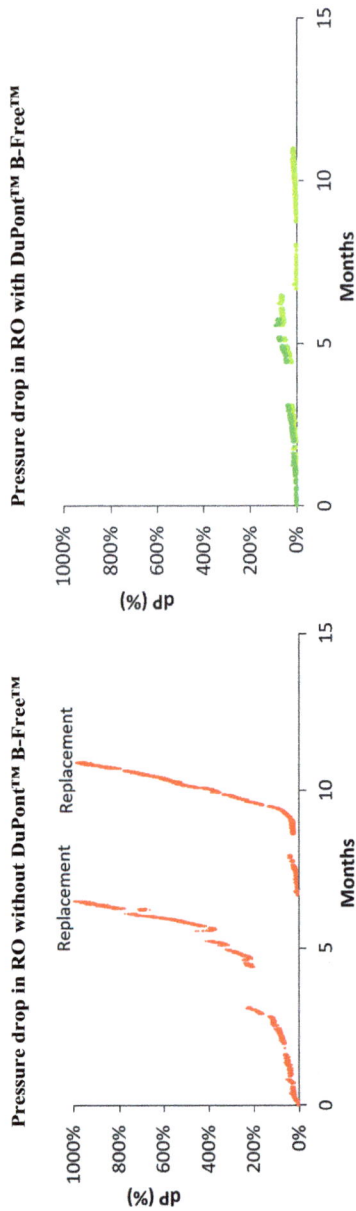

Figure 14.7: Reverse osmosis pressure drop without (left) and with (right) DuPont™ B-Free™.

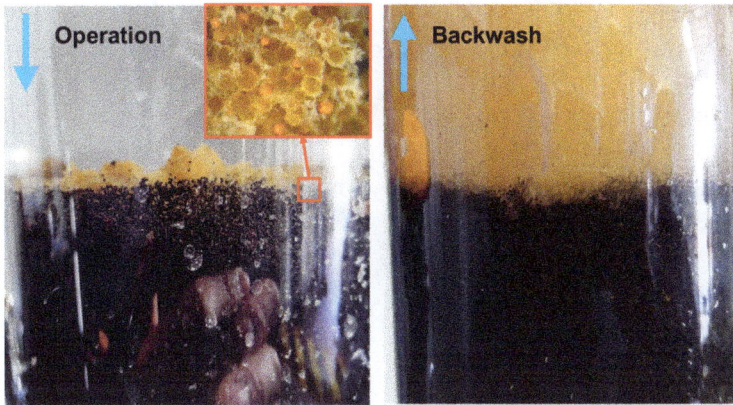

Figure 14.8: Detail of the bio-stratum layer during the backwash process.

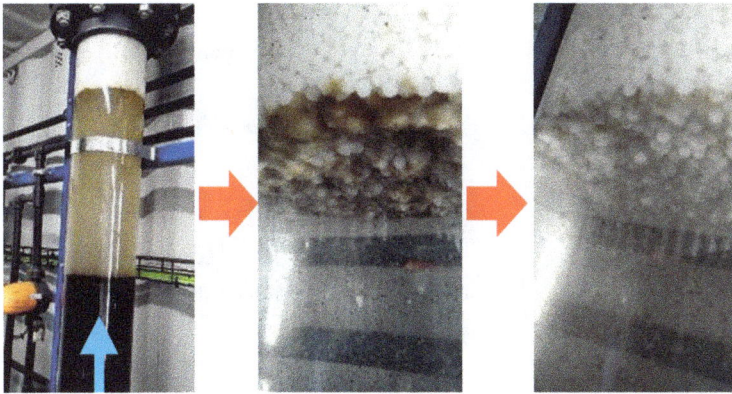

Figure 14.9: Sequence of how the excess biomass exits from the bio-stratum layer through the protective layer on the top of the column during the backwash process.

14.3.5 Organics removal

Since the feed water used was municipal wastewater, it contains a high concentration of organic carbon and nitrogen species. The effect of bio-removal caused by the bio-stratum was quantified by observing the decrease of the total organic carbon (TOC) by 16%, a reduction of the chemical oxygen demand (COD) by 15%, a reduction of the total nitrogen (TN) by 8%, and a reduction of the ammonia (N-NH$_4$) by 1% in the samples after DuPont™ B-Free™ compared to the water samples without it. It should be noted that all the samples were taken at the same time. Figure 14.10 shows that removal rates were quite constant throughout the test. The removal of TN was higher than N-NH$_4$, suggesting that denitrification was dominant, rather than nitrification. COD and TOC removal rates are consistent, showing equivalent removal rates in the range of 15–16%.

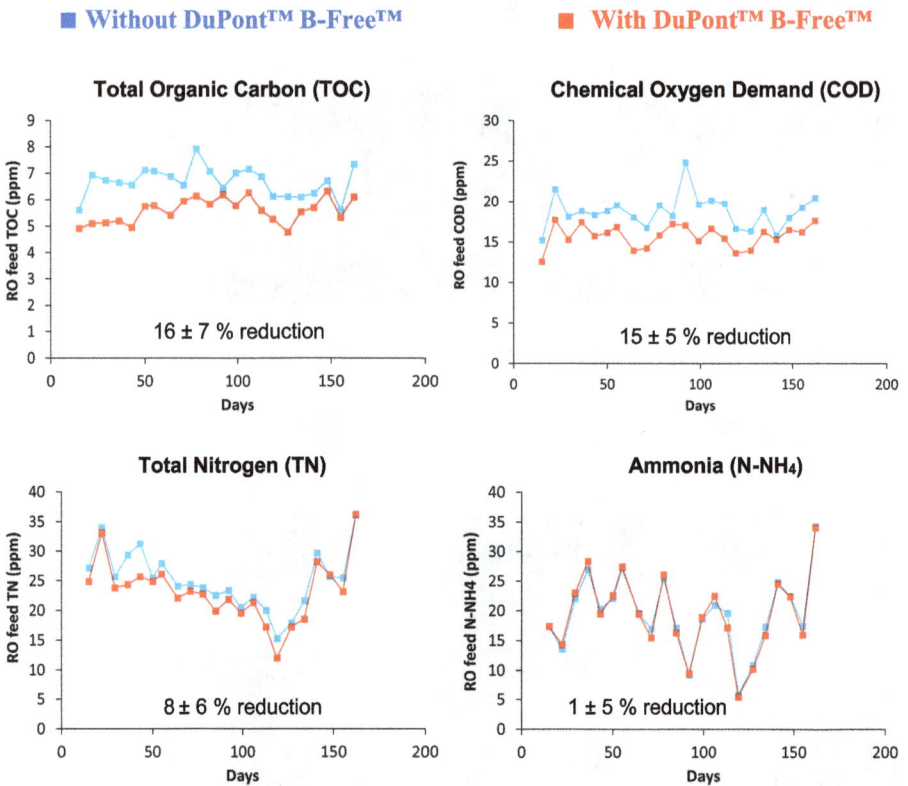

Figure 14.10: Reduction of TOC, COD, TN, and ammonia achieved without (blue) and with (red) DuPont™ B-Free™ pretreatment.

14.3.6 Organics fractions removal

In order to properly identify and quantify the organics present in water, a liquid chromatography-organic carbon detection (LC-OCD) analysis was performed. Water samples were sent to an external laboratory called DOC-Labor GmbH, Germany. This technique allows for the evaluation of the different fractions of dissolved organic carbon (DOC). DOC is composed of hydrophobic dissolved organic carbon (HOC) and hydrophilic dissolved organic carbon (CDOC). Hydrophilic dissolved organic carbon is also referred to as natural organic matter (NOM). NOM is composed of biopolymers, humic substances, building blocks, low molecular weight (LMW) acids, and LMW neutrals.

Each one of these main natural organic matter (NOM) types is explained in more detail below.

The first category corresponds to biopolymers. These show a high molecular weight (> 100 kDa) and are typically hydrophilic compounds. These include polysaccharides and proteinic matter. A sub-fraction of biopolymers called colloidal transparent exopolymer particles is assumed to be responsible for membrane fouling [15].

The second category relates to the humic substances. It should be noted that data for molecular mass and aromaticity are used for this characterization.

The third category relates to building blocks. It assumes the breakdown products of humic substances with molecular weights ranging between 300 and 500 Da. These are the compounds that cannot be removed by flocculation. Also, di- and triprotic acids (oxalic, citric) appear in this fraction.

The fourth category relates to LMW neutrals. These correspond to the complex fraction dominated by weakly or uncharged hydrophilic, or slightly hydrophobic, also called amphiphilic, compounds.

The main idea of this analysis is to understand how much organic matter the Du-Pont™ B-Free™ technology can eliminate. The removal rates of the different categories of dissolved organic matter can be found in Figure 14.11. There, it can be seen how the DuPont™ B-Free™ technology is able to consistently eliminate 18% of the dissolved organic matter. In particular, it can reduce 18% of the hydrophobic dissolved organic matter, and also 18% of the hydrophilic dissolved organic matter, also referred to as NOM. From this NOM, it can reduce 19% of the biopolymers, 30% of the humic substances, 5% of the building blocks, and 9% of the LMW neutrals. It should be noted that the feed absolute units are located in the first white row of the table, while the outlet absolute units are located in the second white row of the table. In gray, one can find the relative percentage of its analyzed compound relative to the rest of the compounds in the same sample, which are either feed or outlet.

14.3.7 Reverse osmosis membranes

The pictures of the two membranes operated without and with DuPont™ B-Free™ pretreatment can be found in Figure 14.12. In these pictures, it can be appreciated how the membrane that was protected with this biofouling prevention technology is not experiencing biofouling, while on the other hand, the membrane that was not protected with DuPont™ B-Free™ technology heavily suffered from biofouling.

NOM

D·O·C LABOR

	Dissolved ppb-C %DOC	Hydrophob. ppb-C %DOC	Hydrophil. ppb-C %DOC	Bio-polymers ppb-C %DOC	DON (Norg) ppb-N	N/C µg/µg	% Proteins in BIOpol %BIOpol	cTEP ppb-C	Humic Subst. (HS) ppb-C %DOC	DON (Norg) ppb-N	N/C µg/µg	Aromaticity (SUVA-HS) L/(mg*m)	Mol.Weight (Mn) g/mol	Position in HS-diagram	Building Blocks ppb-C %DOC	LMW Acids ppb-C %DOC	LMW Neutrals incl. SOM ppb-C %DOC	NO3⁻ ppb-N	NH4⁺ ppb-N	inorg. Colloid. SAC m⁻¹	SUVA (SAC/DOC) L/(mg*m)
Feed	6041	1054 17.4%	4987 82.6%	148 2.5%	17	0,12	35	<3	2323 38.5%	97	0,04	2,33	510	A	1141 18.9%	<3	1374 22.7%	2489	> 6559	< 0,03	3,14
Outlet	4943	860 17.4%	4083 82.6%	120 2.4%	17	0,14	42	<3	1628 32.9%	64	0,04	1,63	476	B	1083 21.9%	<3	1252 25.3%	2190	> 6623	0,05	2,55

Figure 14.11: DuPont™ B-Free™ organic fraction removal based on LC-OCD analysis.

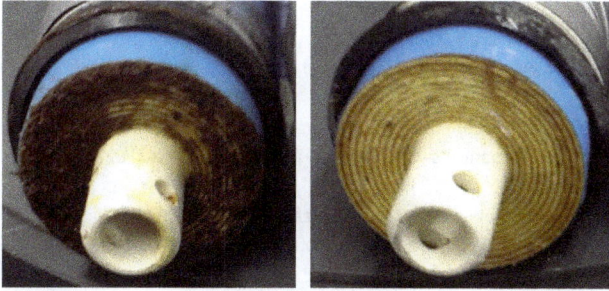

Figure 14.12: Example of two membranes operated without DuPont™ B-Free™ pretreatment (left) and with DuPont™ B-Free™ pretreatment (right).

14.4 Conclusions

DuPont™ B-Free™ pretreatment technology for biofouling prevention in RO is a novel vessel-based media technology that provides instant and sustained biofouling protection. It has been validated using wastewater. For nearly 1 year of operation, it provided stable operation, neutralizing the high biofouling present in the feed water. Moreover, DuPont™ B-Free™ is compact and easy to operate, with only air and water cleaning, eliminating the need for chemicals. The different tests performed with DuPont™ B-Free™ in wastewater demonstrated that the biofouling reduction rate is comparable to what was observed with other types of water, such as river water and seawater.

The required BW rate was, on average, every 5 days, and the higher organic load rate did not cause problems in managing the biomass formation in the DuPont™ B-Free™ column. Additionally, it was shown that DuPont™ B-Free™ extends the lifetime of CFs and RO elements by reducing the fouling rate.

During the pilot trial, an extensive sampling campaign was performed to quantify the carbon and nitrogen bio-removal achieved by the bio-stratum. Nitrate and organic carbon were the main compounds removed from the feed water, which caused lower permeability declines in the RO elements during operation.

References

[1] L. Javier, N. M. Farhat, P. Desmond, R. V. Linares, S. Bucs, J. C. Kruithof, J. S. Vrouwenvelder. Biofouling control by phosphorus limitation strongly depends on the assimilable organic carbon concentration. Water Research, 183 (2020) 116051.

[2] T. Bhattacharjee, S. S. Datta. Confinement and activity regulate bacterial motion in porous media. Soft Matter, 15 (2019) 9920–9930.

[3] M. Schulz, J. Winter, H. Wray, B. Barbeau, P. Bérubé. Biologically active ion exchange (BIEX) for NOM removal and membrane fouling prevention. Water Science and Technology: Water Supply, 17 (2017) 1178–1184.

[4]   H. Yang, B. Minuth, D. Grant Allen. Effects of nitrogen and oxygen on biofilter performance. Journal of the Air & Waste Management Association, 52 (2002) 279–286.

[5]   H. C. Flemming. Biofouling in water systems – Cases, causes and countermeasures. Applied Microbiology and Biotechnology, 76(6) (2007) 1255–1267.

[6]   L. Vanysacker, R. Bernshtein, I. F. J. Vankelecom. Effect of chemical cleaning and membrane aging on membrane biofouling using model organisms with increasing complexity. Journal of Membrane Science, 457 (2014) 19–28.

[7]   Z. Liu, K. M. Lompe, M. Mohseni, P. R. Bérubé, S. Sauvé, B. Barbeau. Biological ion exchange as an alternative to biological activated carbon for drinking water treatment. Water Research, 168 (2020) 115148.

[8]   S. West, H. Horn, W. A. M. Hijnen, C. Castillo, M. Wagner. Confocal laser scanning microscopy as a tool to validate the efficiency of membrane cleaning procedures to remove biofilms. Separation and Purification Technology, 122 (2014) 402–411.

[9]   H. Wray, M. Schulz, J. Winter, P. R. Bérubé, B. Barbeau, W. Bayless, Biological ion exchange for NOM removal – unforeseen synergies. American Water Works Association – Water Quality Technology Conference (WQTC), (2016) 13–17.

[10]   M. Clements, J. Haarhoff. Practical experiences with granular activated carbon (GAC) at the rietvlei water treatment plant. Water SA, 30 (2004) 89–95.

[11]   T. Nguyen, F. A. Roddick, L. Fan. Biofouling of water treatment membranes: A review of the underlying causes, monitoring techniques and control measures. Membranes, 2 (2012) 804–840.

[12]   B. D. Martin, L. De Kock, M. Gallot, E. Guery, S. Stanowski, J. MacAdam, B. Jefferson. Quantifying the performance of a hybrid anion exchanger/adsorbent for phosphorus removal using mass spectrometry coupled with batch kinetic trials. Environmental Technology, 39 (2018) 2304–2314.

[13]   O. Lorain, B. Hersant, F. Persin, A. Grasmick, N. Brunard, J. M. Espenan. Ultrafiltration membrane pre-treatment benefits for reverse osmosis process in seawater desalting. Quantification in terms of capital investment cost and operating cost reduction. Desalination, 203(1–3) (2007) 277–285.

[14]   F. Fatima, S. Fatima, H. Du, R. R. Kommalapati. An evaluation of microfiltration and ultrafiltration pretreatment on the performance of reverse osmosis for recycling poultry slaughterhouse wastewater. Separations, 11(4) (2024) 115.

[15]   M. Monnot, S. Laborie, C. Cabassud. Activated carbon selection for the adsorption of marine DOC and analysis of DOC fractionation. Desalination and Water Treatment, 57(53) (2016) 25435–25449.

Chapter 15
Limiting nutrients to prevent biofouling in seawater: part 1

Industrial-scale pilot at Maspalomas I desalination plant demonstrates the efficiency of DuPont™ B-Free™ pretreatment – a new breakthrough solution against biofouling

Biofouling is one of the most common and severe issues in the operation of seawater reverse osmosis (RO) systems with open intake. If unchecked, it causes significant operational problems such as frequent interruption, damage to the membranes, intense chemical and energy use, and regular cleaning in place (CIP) of the RO membranes. A novel, vessel-based media technology utilized as a membrane pretreatment has been shown to efficiently mitigate the effects of biofouling in RO elements. DuPont™ B-Free™ pretreatment works under different mechanisms which are smartly combined to provide a biostatic environment for downstream RO operations. The Maspalomas I desalination plant with a capacity of 14,500 m^3/day on Gran Canaria Island in Spain has been suffering from biofouling problems in the RO. To resolve the biofouling challenges, experts from Elmasa, a company with more than 45 years of experience in the water industry, collaborated with DuPont Water Solutions and tested for more than a year and a half the novel pretreatment technology – DuPont™ B-Free™ designed to eliminate the effects of biofouling in the RO system. An extensive trial using seawater open intake as source water showed biofouling prevention and trouble-free operation in an industrial-scale pilot plant, while the parallel full-scale plant did continue to suffer from the negative effects of biofouling. DuPont™ B-Free™ creates an instant and sustained biostatic environment for the downstream RO operations and is resilient to upstream upsets.

Acknowledgments: The authors of this chapter would like to acknowledge Elmasa Tecnología del Agua, S.A.U. for their support in this research. Special thanks to Juan Carlos Gonzalez, Sigrid Arenas, Jorge Pordomingo, Imad Kassih, Ruben Mesa and Cristofer Ramos. The author would like to thank all DuPont Water Solutions team for their outstanding expertise in performing these experiments.

This chapter was originally published with the following reference: G. Massons, G. Gilabert-Oriol, S. Arenas-Urrera, J. Pordomingo, J.C. González-Bauzá, E. Gasia, M. Slagt, Industrial-scale pilot at Maspalomas I desalination plant demonstrates the efficiency of DuPont™ B-Free™ pretreatment: a new breakthrough solution against biofouling. Desalination and Water Treatment, 259 (2022) 261–265.

https://doi.org/10.1515/9783111639079-015

15.1 Introduction

15.1.1 Biofouling

Of the various fouling types, biofouling is the most difficult types to manage in reverse osmosis (RO) systems [1]. Commercial plants are not sterile environments and any microorganism that enters the system will rapidly multiply. Microorganisms tend to attach and generate biofilms on the membrane, feed spacer, and, to a lesser extent, on the membrane surface. The formation of bacterial biofilms is initiated with the attachment of planktonic, free-swimming bacteria on the surface. The biofilm matrix is a sticky polymeric structure used by microorganisms to immobilize themselves and grow on a surface [2]. The main component of the matrix is a strongly hydrated mixture of carbohydrates and proteins, known as extracellular polymeric substances (EPS). EPS stabilize biofilms by holding the microbial cells together and attaching the growing biofilm to the membrane and the feed spacer surface, creating structures that gradually clog the feed-concentrate channel [3]. Thus, the void space of a feed-concentrate channel is reduced and the resistance of water to flow increases. This effect is associated with an exponential pressure drop increase [4]. Biofilms reduce the RO module efficiency and increase their energy consumption. Additionally, biofilms are challenging to clean due to the sticky nature of EPS, which provides high mechanical and chemical stability to bacteria [5].

Despite multiple attempts to tackle biofouling, it still remains a key unsolved problem in the water treatment industry [6]. It is currently reported that around 70% of the RO water treatment plants in the Middle East are experiencing problems with biofouling [7]. Additionally, it is also reported that around 83% of all the surface water plants in the United States are reported to have problems with biofouling [8].

Biofouling still remains as the key challenge to address in the RO membrane systems. In particular, a comprehensive study done by Genesys, a company specialized in performing autopsies to the RO membranes, after studying almost 100 RO seawater membranes that they autopsied and analyzed between 2002 and 2010, revealed that among all the symptoms that those membranes exihibited, the most prevalent case among all was biofouling. Thus, biofouling was identified as the number one reason as why membranes were failing. And, in particular, biofouling showed to be prevalent with 27% of all the cases analyzed [9]. In order to corroborate these findings and understand if, even one decade later, biofouling was still the main reason that accounted for RO membrane failure, different customers were interviewed globally by DuPont Water Solutions in 2019. As a result of these interviews, it was observed that biofouling was still the key reason for RO membrane failure. In particular, it was shown as the main reason for membranes to fail, with a 32% of the cases being accounted because of biofouling [10].

Biofouling presents a key challenge in an RO membrane system. As biofouling leads to a pressure drop increase, this results in higher energy consumption. This increase in differential pressure later necessitates performing a chemical cleaning (CIP).

This involves using chemicals in an attempt to restore the initial performance of the RO membrane. However, each subsequent cleaning leads to a decrease in uptime, as the installation needs to be stopped in order to clean the membrane. Biofouling can also lead to a deterioration in the permeate water quality. This phenomenon is referred to as the biofilm enhanced osmotic pressure. This happens as the biofilm accumulation on the membrane surface tends to accumulate salt on it, thus leading to an increase in the concentration polarization, which leads to a higher salt passage across the membrane [11, 12]. Eventually, after multiple chemical cleanings on a membrane, and after repeated exposure to a growing biofilm, the membrane can experience a loss in its performance or some mechanical damage, which can lead to a premature replacement of the RO membrane, thus shortening its lifetime. All this can significantly affect the total cost of water in an RO water treatment plant.

15.1.2 Method to mitigate biofouling

The sensitivity of polyamide-based membranes to oxidizing agents, such as chlorine, greatly limits the use of chemicals to prevent bacterial growth in the feed water [13]. Additionally, traditional pretreatments such as coagulation, flocculation, ultrafiltration, or cartridge filters are not effective in removing the biofouling potential of the feed water [14].

The buildup of biofilm on RO systems is not only challenging to prevent but also very difficult to remove once established [15]. Caustic solutions can potentially hydrolyze the polysaccharides and proteins from the EPSs matrix to disperse the fouling layer. However, cleaning methods for biofouling are not efficient enough [16]. Even harsh cleaning conditions cannot fully remove biofouling from RO modules, resulting in rapid regrowth after each CIP [17].

Another alternative is the metabolic inactivation of bacteria that can be applied before water reaches the RO modules, minimizing the risk of bacteria regrowth. The inactivation can be via physical, such as ultraviolet light, or chemical, such as with biocides [18]. Biocides have been recognized as efficient compounds to prevent biofouling formation. Unlike physical inactivation, biocides are dissolved in the feed water and are effective throughout the system. Many of the commonly used biocides are oxidizing in nature and would damage the polyamide membrane over time, such as with sodium hypochlorite or chloramines. Only the nonoxidizing biocides might be partially compatible with thin-film composite RO membranes. However, the severe environmental and health hazards that biocides impose must be carefully evaluated when considering implementation.

15.1.3 Biofouling prevention technology

DuPont™ B-Free™ aims to avoid the drawbacks of current methods to prevent RO bio-fouling, being a compact, efficient, chemical–free, and robust technology. This pre-treatment technology is vessel-based and eliminates the detrimental effects of biofouling. It creates an instant and sustained biostatic environment [19] for the downstream RO operations without the need to use chemicals during operations and is resilient to process upsets. This protection is created by three different media each having a specific purpose in the process, as shown in Figure 15.1.

The idea of DuPont™ B-Free™ technology is inspired in the concept described in this book, where it is observed that biofouling mainly develops on the lead elements. Additionally, it is also observed that biofouling needs nutrients in order to develop, and more specifically, it needs easily assimilable nutrients. In this book, the importance of dosing assimilable nutrients in order to quickly promote biofouling has been described. This biofouling prevention technology uses these two concepts in a novel approach, with the idea that biofouling is growing in the wrong place. So, instead of making it grow in the RO membrane, where it is very challenging to get it cleaned, it grows in the DuPont™ B-Free™ technology, where it is easy to control its growth and maintain it in a well-controlled place. In order to clean it, only water and air are needed. As DuPont™ B-Free™ technology uses spherical media of 750–950 μm, it is very easy to clean as it is completely mobile as opposed to cleaning a feed spacer that is fixed and trapped into an RO membrane.

Therefore, DuPont™ B-Free™ technology uses three well-established and synergistic mechanisms to create a downstream biostatic environment so that it remains very challenging for RO to develop biofouling.

The first mechanism happens in the bio-stratum layer. This layer contains spherical media between 750 and 950 μm and can be around 30 cm thick. It is designed so that bacteria can adhere in the free-void spaces between the spherical media, colonize it, and start building a biomass. This biomass acts as a biological filter that removes the majority of assimilable nutrients. Additionally, this bio-stratum layer acts as a sieving mechanism so that the majority of total suspended solids (TSS) can be removed, in case there are any present. These two mechanisms contribute to creating a downstream biostatic environment so that it is more difficult for the RO located after this pretreatment to grow biofouling.

The second mechanism happens in the Safeguard layer. This layer contains spherical media between 300 and 1,200 μm and can be around 50 cm thick. It is designed so that any phosphate that has not been digested in the previous Bio-stratum layer can be absorbed and eliminated to this Safeguard layer. This phosphate polishing step contributes to make the system more reliable, as it helps ensure the downstream water does not have the potential to build biofouling in the RO system. This is especially important during start up, during changes in feed water composition, or after a chemical cleaning, when the biomass might not be properly established in the bio-stratum layer. This layer also helps filtrating and retaining any TSS that might leak, or any biomass carry over that might accidentally go to the downstream RO and eventually contribute to developing biofouling.

It should be noted that between the bio-stratum layer and the Safeguard layer, there is a distributor. This distributor is placed between both layers so that any maintenance backwash can be done effectively and prevent any particle coming from the biomass after a backwash can be migrated from the bio-stratum layer into the safeguard layer. Additionally, by placing this intermediate distributor, energy can be saved during the backwash and chemical cleanings. It should be noted that it is also possible to perform a chemical cleaning (CIP) using a caustic solution.

Finally, the third mechanism happens in the protective layer. This layer contains spherical media between 2.5 and 4 cm and can be around 15–30 cm thick. This media is a floating inert material which is located on top of the vessel, wrapping up the distributors. It is designed so that during the backwash sequence, the biomass can easily exit the vessel, and no media is lost during the backwash flow. This effective separation and filtration of the biomass from the media help sustain the operation of the DuPont™ B-Free™ technology.

Additionally, it should be noted that a free space, called Freeboard, needs to be left between the Protective Layer and the bio-stratum. This free space is needed so that during the backwash, the media can be properly fluidized, and the biomass can

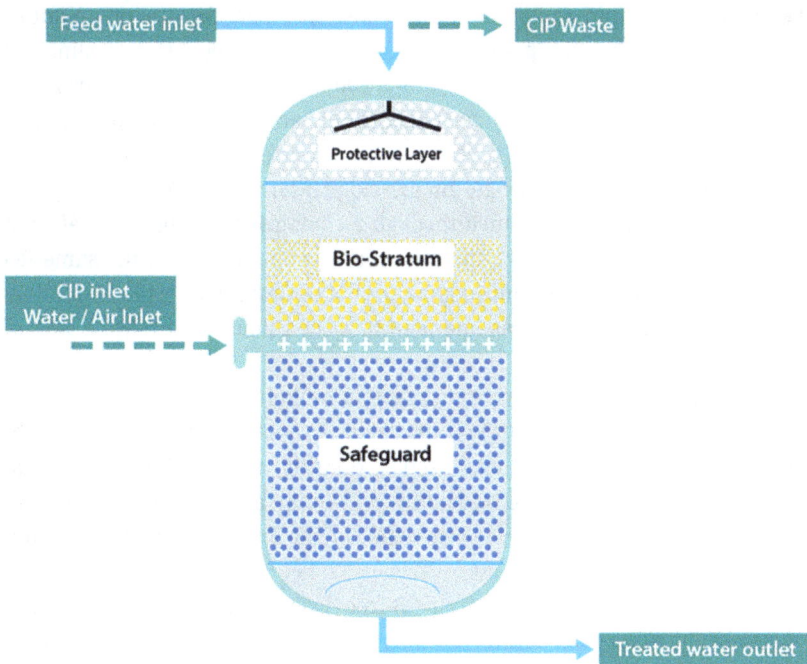

Figure 15.1: DuPont™ B-Free™ vessel layout.

be effectively separated from the media. This Freeboard can have a space of at least the same height as the bio-stratum layer, and ideally, an additional 30% or more height than the safeguard has.

15.2 Materials and methods

15.2.1 Experimental plant

The study was conducted in the Maspalomas I desalination plant located in the Bahía Feliz region in San Bartolomé de Tirajana, Gran Canaria, Spain. This desalinaiton plant is located to the east of Maspalomas municipality. The location of the plant as well as a bird's eye view photo of it can be found in Figure 15.2.

The Maspalomas I desalination plant with a capacity of 14,500 m³/day is located on Gran Canaria Island in Spain. It has been suffering from biofouling problems in the RO system. To resolve the biofouling challenges, experts from Elmasa Tecnología del Agua, a company with more than 45 years of experience in the water industry, collaborated with DuPont Water Solutions and tested for more than a year and a half a novel pretreatment technology, the DuPont™ B-Free™ designed to eliminate the effects of biofouling in the RO system.

DuPont™ B-Free™ biofouling prevention performance was evaluated by comparing an RO element installed in a pilot plant to RO elements from desalination plant as a control. A schematic of the pilot plant can be found in Figure 15.3. The desalination plant has an open seawater intake, followed by a 100-µm screen filter, an ultrafiltration pretreatment, followed by a 1-µm cartridge filter, and the two RO passes. The pilot plant is set up in parallel after the ultrafiltration pretreatment. It consists of the DuPont™ B-Free™ system, followed by a 1-µm cartridge filter and the RO line.

This membrane element was cleaned through a chemical cleaning (CIP) before it was used for this side-by-side testing. The pilot plant is operated with the same RO type, feed water, and operating conditions as the large desalination plant to allow comparison. In particular, the RO was operated at a feed flow of 8.1 m³/h, a flux of 24 L/m²h, and a recovery of 11%.

A constant pressure drop over time indicates that biofilm is not developing within the RO module. The desalination plant was operated without DuPont™ B-Free™ technology, and its pressure drop trend was used to quantify the number of chemical cleanings prevented and to assess the biofouling prevention achieved by the pretreatment. After operation, a detailed visual inspection of the RO was completed to confirm operational results in the pilot plant.

The vessel layout used to operate DuPont™ B-free™ is presented in Figure 15.4. The inner diameter of the glass fiber-reinforced polymer composites vessel is 1,000 mm, and given the feed flow of 17.7 m³/h (77.9 gpm), the linear operating velocity is 22.5 m/h. The bed height is 80 cm, corresponding to 630 L of media. During the period of test

Figure 15.2: Maspalomas I desalination plant location in Gran Canaria.

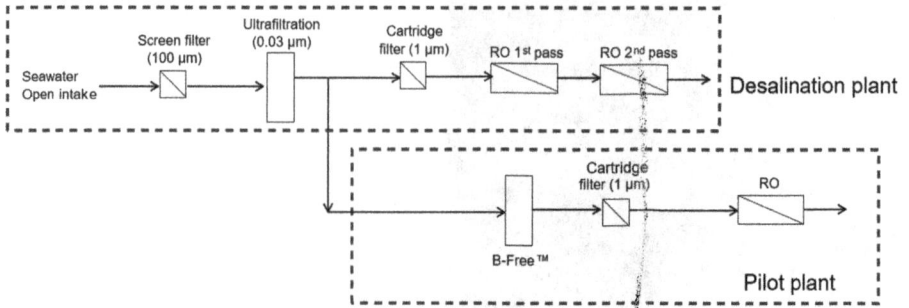

Figure 15.3: Pilot plant setup with DuPont™ B-Free™ technology.

presented in this report, it is operated at 28 BV/h. Feed water port (A) contains lateral pipes with wedge wires of 1-mm slot size used for feed water and to collect waste back-wash water. Intermediate port (B) contains lateral pipes with wedge wires of 0.2-mm slot size and is used for the regular backwash process, injecting air and water to clean the media. Air flow during backwash is 60 m³/h and backwash (BW) water flow is 10 m³/h. Treated water port (C) contains a 0.2 nozzle plate (46 nozzles/m²) and is used to collect the filtrated water and eventually drain the vessel.

15.2.2 Seawater characteristics

Feed water to the desalination plant has undergone an extensive pretreatment, but despite salinity, low total organic carbon load, and low phosphate concentration, as summarized in Table 15.1, it still has an intrinsically high biofouling potential, leading to a rapid pressure drop buildup.

15.3 Results and discussion

15.3.1 Backwash

During the whole operation period, several backwashes were performed to restore the pressure drop of the DuPont™ B-Free™ system. As can be observed in Figure 15.5, the backwash protocol with water and air was effective in restoring performance without the need to use chemical cleanings. The pressure drop (dP) trigger used was around 0.7 bar. As it can be seen from the figure, during the whole operating time, only eight backwashes were needed to remove the excess biomass in the DuPont™ B-Free™ vessel. It is also worth pointing out that the backwashes always enabled the restoration of the pressure drop to its initial level. This ensures that the system can operate in a reliable and sustainable way, and no irreversible biomass accumulates in

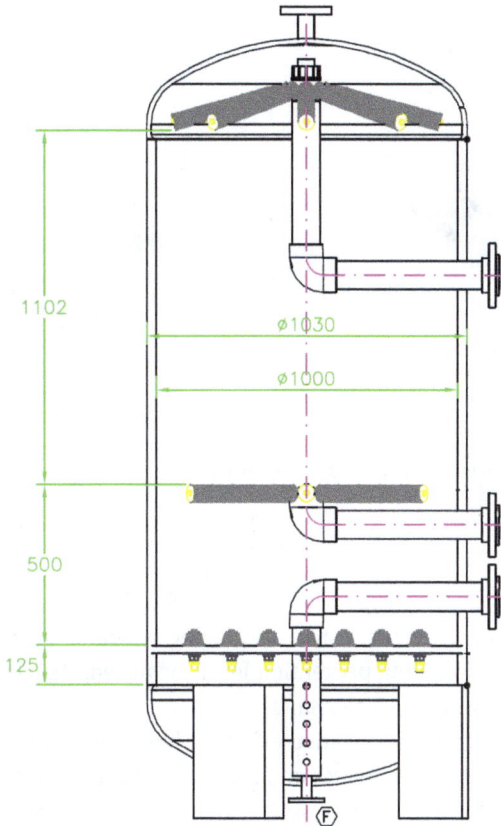

Figure 15.4: DuPont™ B-Free™ vessel layout.

Table 15.1: Feed water properties.

Parameter	Value
Source water	Seawater open intake
Plant capacity	17,700 L/h (77.9 gpm)
Conductivity (µS/cm)	56,362
Temperature (°C)	20–23 (68–74 F°)
pH	8.02
TOC (mg/L)	1.28
PO$_4$ (µg/L)	40

the vessel. This led to a water recovery of 99.95% for the whole operating period. If time is taken into account, this ensured a time availability of 99.85% that the DuPont™ B-Free™ vessel stayed online. Defining the vessel efficiency as the product of water recovery and availability, this led to a vessel efficiency of 99.80%.

Pressure drop in DuPont™ B-Free™ vessel

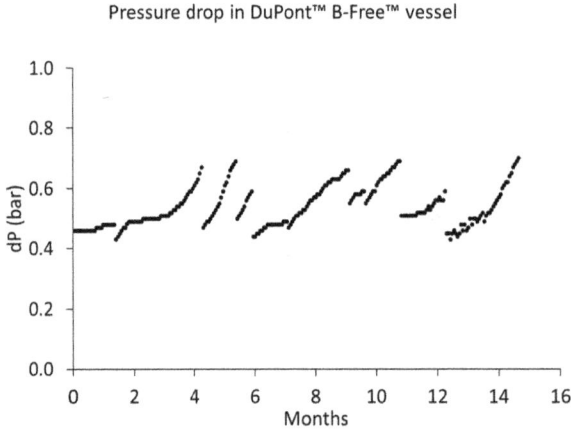

Figure 15.5: Backwash overview during operation in the DuPont™ B-Free™ vessel.

15.3.2 Cartridge filters

The extensive trial has lasted more than 1.5 years. It has been using seawater open intake as source water highlighted the biofouling prevention levels achieved, thanks to DuPont™ B-Free™ for the cartridge filter and RO elements.

The main plant continued to suffer from the negative effects of biofouling. However, the pretreatment installed in the pilot plant creates an instant and sustained biostatic environment for the downstream RO operations and is resilient to upstream upsets.

This can be quantified for the cartridge filter replacement rate, with 12 replacements in the large plant without pretreatment, compared to three replacements in the pilot plant. These trends can be observed in Figure 15.6.

15.3.3 Reverse osmosis

Looking at the RO operation, seven chemical cleanings were required during the testing period compared with the stable operation with no chemical cleaning in the RO membranes operated with DuPont™ B-Free™. These evolutions can be seen in Figure 15.7.

15.3.4 Bio-stratum growth

In order to understand how the biomass was distributed across the media present in the vessel, different samples of the media were taken at 5, 44, and 69 cm depth, from a total of 80 cm depth, just before performing a maintenance backwash. From these samples, it

Pressure drop in CF without DuPont™ B-Free™

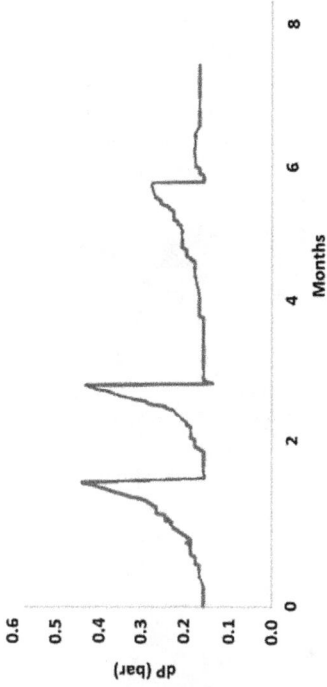

Pressure drop in CF with DuPont™ B-Free™

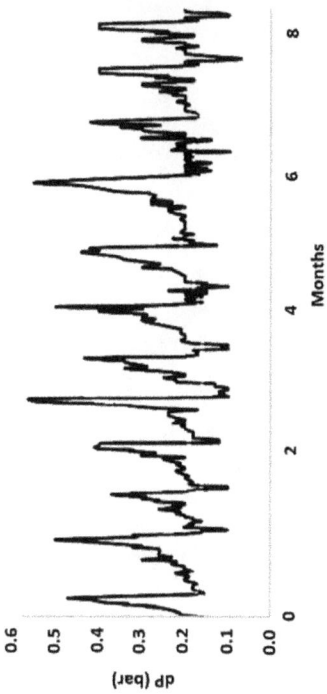

Figure 15.6: Cartridge filter pressure drop evolution in seawater test.

Pressure drop in RO without DuPont™ B-Free™

Pressure drop in RO unit with DuPont™ B-Free™

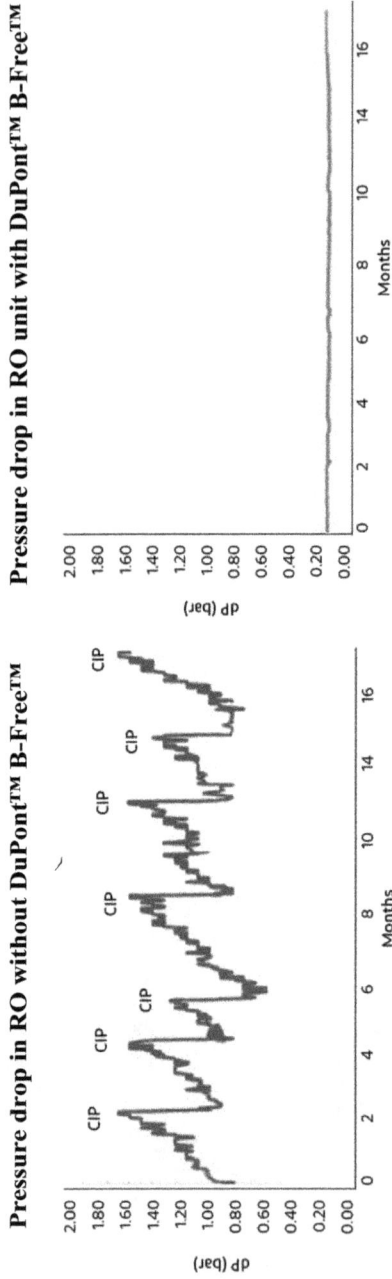

Figure 15.7: Pressure drop evolution in the reverse osmosis system not protected (left) and protected (right) with DuPont™ B-Free™ technology.

could be seen that biomass was mainly present on the top layer, as it can be observed in Figure 15.8. The samples taken at 44 and 69 cm depth hardly had any biomass.

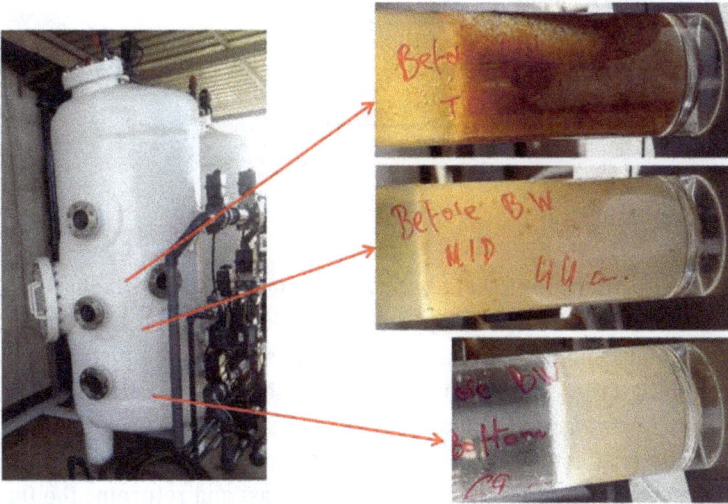

Figure 15.8: Biomass distribution in the DuPont™ B-Free™ vessel.

After the backwash was completed, samples at the same depths used previously were obtained from the bed. Figure 15.9 shows how backwash effectively removed excess biomass formation on top media layer without causing contamination of deeper bed areas.

Figure 15.9: Biomass stratification in media bed after backwash.

Each sample was analyzed, and as it can be seen in Table 15.2, it can be confirmed that the main biomass is accumulated on the top of the media.

Table 15.2: Biomass concentration found in different media depth.

Sample	TOC (mg/L)	TN (mg/L)	Proteins (mg/L)	Carbohydrates (mg/L)
Top – 5 cm	17.6	13.4	3.6	5.0
Mid – 44 cm	8.6	5.1	0.7	1.8
Bottom – 69 cm	8.5	6.8	0.2	0.5

15.3.5 Backwash observation

In order to investigate further the effect of backwash into removing the biomass accumulated in the DuPont™ B-Free™ vessel, photos were taken of the bio-stratum top before and after the maintenance backwash with an air scour. These pictures can be found in Figure 15.10. From these pictures, it can be seen that before the backwash, the bio-stratum layer was full of biomass. Then, it can be observed that after the backwash, biomass is effectively removed. This showcases how effective is the backwash maintenance sequence in removing excess biomass and returning the DuPont™ B-Free™ system to its original state, so it can keep preventing biofouling into its downstream RO system, or any other system that might lay downstream that can be prone to biofouling.

Before backwash   **After backwash**

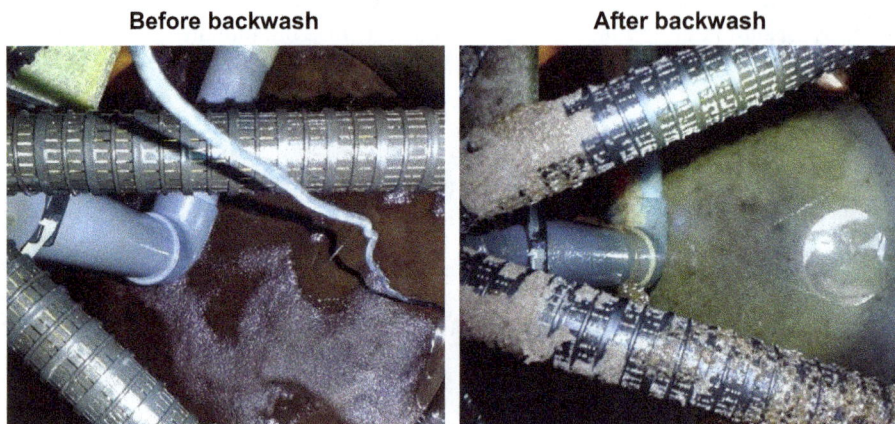

Figure 15.10: Backwash effect on the biomass present on the bio-stratum layer.

15.3.6 Membrane autopsies results

The piloted elements were removed and autopsied after the test. As a visual aid, the autopsy of an 8-in membrane element biofouled from another trial is provided since the control element from the tests could not be autopsied. Figure 15.11 showed very little biofouling presence on the membrane surface despite operating more than 1.5 years.

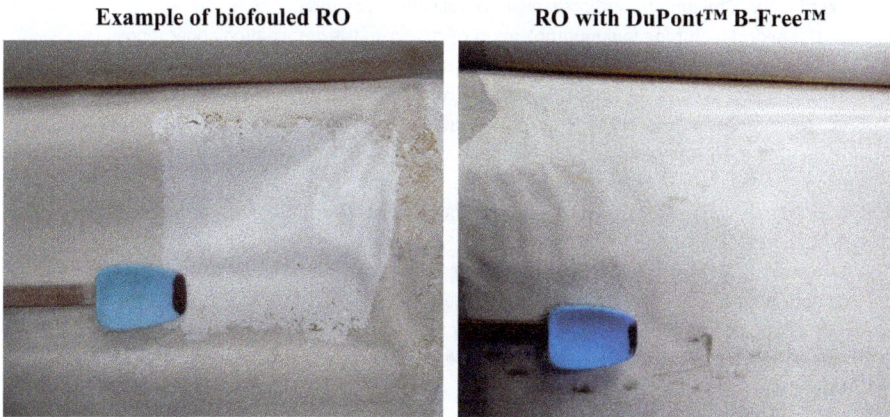

Figure 15.11: Elements autopsy showing how the membranes pretreated with B-Free™ technology mitigated biofouling.

15.4 Conclusions

This pretreatment technology for biofouling prevention in RO is a novel vessel-based media technology that provides instant and sustained biofouling protection. It has been validated using seawater at a commercial desalination plant with industrial-sized elements. For more than 1.5 years of piloting, it provided stable operation minimizing overall system downtime.

DuPont™ B-Free™ is compact and easy to operate, with cleaning performed with only air and water, eliminating the need of chemicals. Additionally, there are no special requirements to upgrade existing systems.

References

[1] L. Javier, N. M. Farhat, P. Desmond, R. V. Linares, S. Bucs, J. C. Kruithof, J. S. Vrouwenvelder. Biofouling control by phosphorus limitation strongly depends on the assimilable organic carbon concentration. Water Research, 183 (2020) 116051.

[2] T. Bhattacharjee, S. S. Datta. Confinement and activity regulate bacterial motion in porous media. Soft Matter, 15 (2019) 9920–9930.

[3]   M. Schulz, J. Winter, H. Wray, B. Barbeau, P. Bérubé. Biologically active ion exchange (BIEX) for NOM removal and membrane fouling prevention. Water Science and Technology: Water Supply, 17 (2017) 1178–1184.

[4]   H. Yang, B. Minuth, D. Grant Allen. Effects of nitrogen and oxygen on biofilter performance. Journal of the Air & Waste Management Association, 52 (2002) 279–286.

[5]   H. C. Flemming. Biofouling in water systems – Cases, causes and countermeasures. Applied Microbiology and Biotechnology, 76(6) (2007) 1255–1267.

[6]   A. Matin, Z. Khan, S. M. J. Zaidi, M. C. Boyce. Biofouling in reverse osmosis membranes for seawater desalination: Phenomena and prevention. Desalination, 281 (2011) 1–16.

[7]   M. G. Khedr. Membrane fouling problems in reverse-osmosis desalination applications. International Desalination and Water Reuse Quarterly, 10(3) (2000) 8–17.

[8]   S. P. Chesters, N. Pena, S. Gallego, M. Fazel, M. W. Armstrong, F. Del Vigo. Results from 99 seawater RO membrane autopsies. IDA Journal of Desalination and Water Reuse, 5(1) (2013) 40–47.

[9]   DuPont, Interviewing customers on the importance of biofouling, DuPont Water Solutions Internal Report, (2019)

[10]   D. H. Paul. Reverse osmosis: Scaling, fouling and chemical attack. Desalination Water Reuse, 1 (1991) 8–11.

[11]   M. Herzberg, M. Elimelech. Biofouling of reverse osmosis membranes: Role of biofilm-enhanced osmotic pressure. Journal of Membrane Science, 295(1–2) (2007) 11–20.

[12]   T. H. Chong, F. S. Wong, A. G. Fane. The effect of imposed flux on biofouling in reverse osmosis: Role of concentration polarisation and biofilm enhanced osmotic pressure phenomena. Journal of Membrane Science, 325(2) (2008) 840–850.

[13]   L. Vanysacker, R. Bernshtein, I. F. J. Vankelecom. Effect of chemical cleaning and membrane aging on membrane biofouling using model organisms with increasing complexity. Journal of Membrane Science, 457 (2014) 19–28.

[14]   Z. Liu, K. M. Lompe, M. Mohseni, P. R. Bérubé, S. Sauvé, B. Barbeau. Biological ion exchange as an alternative to biological activated carbon for drinking water treatment. Water Research, 168 (2020) 115148.

[15]   S. West, H. Horn, W. A. M. Hijnen, C. Castillo, M. Wagner. Confocal laser scanning microscopy as a tool to validate the efficiency of membrane cleaning procedures to remove biofilms. Separation and Purification Technology, 122 (2014) 402–411.

[16]   H. Wray, M. Schulz, J. Winter, P. R. Bérubé, B. Barbeau, W. Bayless, Biological ion exchange for NOM removal – unforeseen synergies. American Water Works Association – Water Quality Technology Conference (WQTC), (2016) 13–17

[17]   M. Clements, J. Haarhoff. Practical experiences with granular activated carbon (GAC) at the rietvlei water treatment plant. Water SA, 30 (2004) 89–95.

[18]   T. Nguyen, F. A. Roddick, L. Fan. Biofouling of water treatment membranes: A review of the underlying causes, monitoring techniques and control measures. Membranes, 2 (2012) 804–840.

[19]   B. D. Martin, L. De Kock, M. Gallot, E. Guery, S. Stanowski, J. MacAdam, B. Jefferson. Quantifying the performance of a hybrid anion exchanger/adsorbent for phosphorus removal using mass spectrometry coupled with batch kinetic trials. Environmental Technology, 39 (2018) 2304–2314.

Chapter 16
Limiting nutrients to prevent biofouling in seawater: part 2

Testing of DuPont™ B-Free™ technology in Arabic Gulf water at Sharjah Electricity & Water Authority (SEWA) Hamriyah Desalination Plant

Increasing water and power demand in the Middle East requires that desalination plants maximize overall efficiency and operate at the lowest production cost. Despite the extended pretreatment in the Hamriyah plant, the deteriorating quality of seawater over the last few years, heavy marine biological growth, and the extended red tide have imposed serious challenges for seawater reverse osmosis (SWRO) operation. Sharjah Electricity & Water Authority aims to achieve the best long-term performance and reliability to maximize plant availability throughout the year. DuPont™ B-Free™ creates an instant and sustained biostatic environment for the downstream reverse osmosis operations and is resilient to upstream upsets. This pretreatment will allow water production to be maintained during all regular variations in seawater quality throughout the seasons and in future. A demo unit was installed and commissioned onsite to demonstrate the reduction in fouling potential achieved, thanks to DuPont™ B-Free™ pretreatment technology. This long-term test has shown significant improvements in SWRO operation and the mitigation of biofouling potential. These results are translated into reduced energy consumption, reduced chemical cleaning, and upgraded system reliability.

16.1 Introduction

16.1.1 Biofouling

The management of biofouling in reverse osmosis (RO) systems presents a significant challenge due to the rapid proliferation of microorganisms in the typically nonsterile environments of commercial facilities [1]. These microorganisms predominantly adhere to the membrane feed spacer and, to a lesser extent, the membrane surface,

Acknowledgments: The author would like to acknowledge the contributors to this chapter: Gerard Massons, Guillem Gilabert-Oriol, Marc Slagt, Rajesh Balakrishnan, Hardik Pandya, Alaa Elsayed, and Harith Alomar. The author would like to thank the entire DuPont Water Solutions team for their outstanding expertise in performing these experiments.

This chapter was originally published with the following reference: G. Massons, G. Gilabert-Oriol, M. Slagt, R. Balakrishnan, H. Pandya, A. Elsayed, H. Alomar, Testing of DuPont™ B-Free™ technology in Arabic Gulf water at Sharjah Electricity & Water Authority (SEWA) Hamriyah Desalination Plant. Desalination and Water Treatment, 309 (2023) 80–83.

https://doi.org/10.1515/9783111639079-016

where they form biofilms. Biofilm formation is initiated when free-floating bacteria settle on the surface and begin to multiply [2]. The biofilm is held together by extracellular polymeric substances (EPS), a sticky, hydrated mixture of proteins and carbohydrates that provides structural support to the growing microbial community [3]. EPS facilitates the attachment of microbial cells to both the membrane and feed spacer, which accelerates the biofilm's growth and leads to the gradual narrowing of the feed-concentrate channel. This blockage increases flow resistance and leads to an exponential increase in pressure drop (dP) [4], thereby decreasing the efficiency of the RO system and raising energy costs. Moreover, EPS provides the biofilm with strong mechanical and chemical stability, making it extremely difficult to clean [5].

16.1.2 Methods to mitigate biofouling

Polyamide-based membranes are vulnerable to oxidation from agents like chlorine, making it difficult to use certain chemicals to control bacterial growth in feed water [6]. Moreover, conventional pretreatment processes, including coagulation, flocculation, ultrafiltration, and cartridge filtration, do not effectively address the potential for biofouling in feed water [7]. Once biofilms form on the surfaces of RO systems, preventing and eliminating them becomes a particularly challenging task [8]. Chemical cleaning agents can break down the polysaccharides and proteins in the EPS matrix, allowing the fouling layers to be dispersed. However, the effectiveness of these methods remains limited, particularly under challenging conditions [9]. Furthermore, aggressive cleaning procedures are not enough to prevent biofilm regrowth and often lead to quick reformation after each chemical cleaning (CIP) [10].

An alternative strategy to mitigate bacterial growth is metabolic inactivation, which involves treating the water in-line before it reaches the RO modules, thus reducing the likelihood of biofilm reformation. There are two key inactivation methods: physical and chemical [11]. Biocides can be effective at preventing biofouling formation, but many commonly used biocides, such as hypochlorite and chloramines, are oxidizing agents that degrade polyamide membranes over time. Only nonoxidizing biocides can effectively control bacterial activity without causing damage to thin-film composite membranes. However, it is crucial to consider the potential environmental and health risks associated with biocidal agents when deciding whether to use them.

16.1.3 Biofouling prevention technology

DuPont™ B-Free™ is an innovative pretreatment technology addressing the limitations of current methods for mitigating RO biofouling. By creating a biostatic environment for downstream RO operations, it offers a compact, chemical-free, and robust solution [12]. This vessel-based system employs three synergistic mechanisms across distinct media layers, designed for easy maintenance using only water and air, as shown in Figure 16.1.

The development of DuPont™ B-Free™ builds on insights from the previous chapters of this book, where the mechanisms of nutrient removal and the role of suspended solids in biofouling were extensively analyzed. These earlier chapters underscored how nutrient-rich feedwater and insufficient filtration exacerbate fouling risks. DuPont™ B-Free™ incorporates these findings by implementing a process that actively removes assimilable nutrients and filters solids upstream, significantly reducing fouling potential.

The bio-stratum layer (750–950 µm, 30 cm thick) promotes bacterial adhesion and biomass formation in controlled void spaces, acting as both a biological filter and a sieve for total suspended solids (TSS).

The safeguard layer (300–1,200 µm, 50 cm thick) provides phosphate polishing and retains residual solids or biomass, ensuring consistency during operational variations or after chemical cleanings. An intermediate distributor enhances backwash efficiency and protects the safeguard layer.

The protective layer (2.5–4 cm, 15–30 cm thick) prevents media loss during backwashing, supported by a freeboard zone allowing for proper media fluidization. Together, these layers work to sustain biofouling control while reducing operational complexity and environmental impact.

By leveraging the findings presented in the previous chapter, DuPont™ B-Free™ exemplifies how integrating nutrient control and filtration mechanisms upstream can effectively prevent RO biofouling.

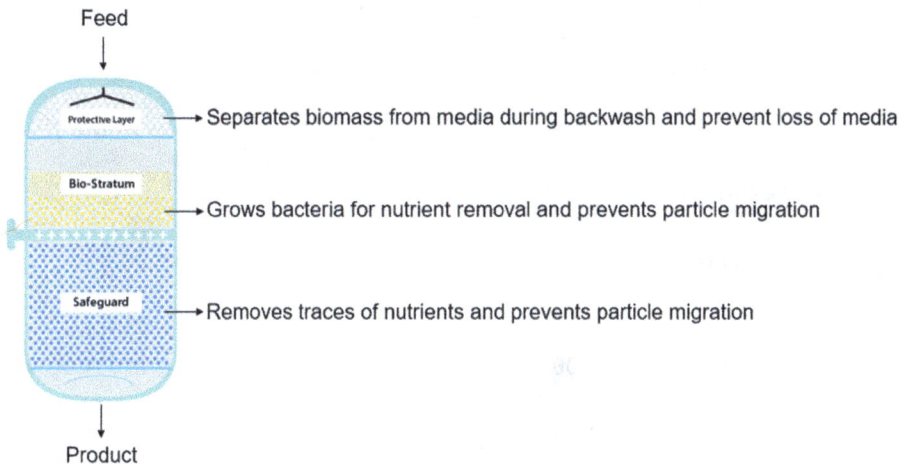

Feed

Protective Layer → Separates biomass from media during backwash and prevent loss of media

Bio-Stratum → Grows bacteria for nutrient removal and prevents particle migration

Safeguard → Removes traces of nutrients and prevents particle migration

Product

Figure 16.1: DuPont™ B-Free™ vessel layout.

The primary objective of biofouling prevention technologies, such as DuPont™ B-Free™, is to address different types of fouling effectively. For example, ultrafiltration pretreatment targets TSS, efficiently removing them to enhance the performance, efficiency, and reliability of downstream RO systems [13, 14]. Similarly, the DuPont™ B-Free™ technology focuses on managing biofouling, allowing the RO system to concentrate on its core function, which is removing ^total dissolved salts (TDS) from water at optimal efficiency. By integrating ultrafiltration, biofouling prevention, and RO, this combined approach maximizes system efficiency and reliability, minimizing particulate and biofouling issues in RO operations. Figure 16.2 illustrates this synergy among the three technologies.

As discussed in the previous chapters, understanding the challenges of fouling mechanisms lays the foundation for appreciating how technologies like DuPont™ B-Free™ contribute to system optimization.

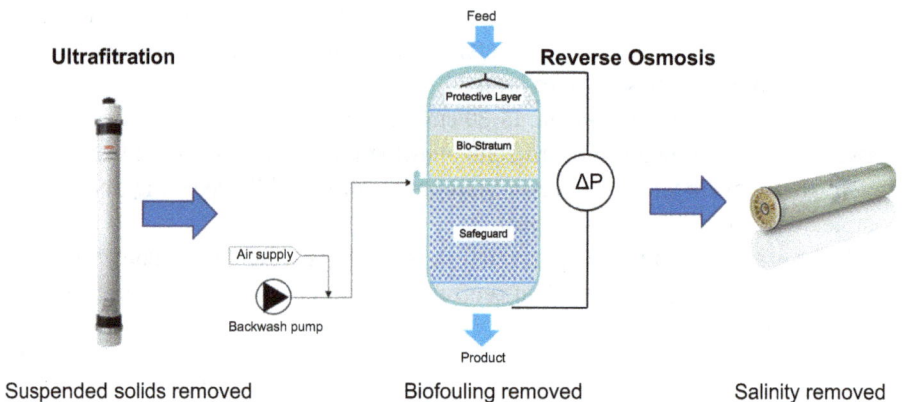

Figure 16.2: DuPont™ B-Free™ system integrated between the ultrafiltration and reverse osmosis systems.

16.2 Materials and methods

16.2.1 Experimental plant

The study was conducted at the Sharjah Electricity & Water Authority (SEWA) Hamriyah desalination plant (90,000 m^3/d) located in the Sharjah region in the United Arab Emirates. It uses ultrafiltrated sea water from an open intake as a feed source and has been suffering from biofouling problems in the RO system. To resolve the biofouling challenges, a pilot plant was deployed to test the novel pretreatment technology. This technology is called DuPont™ B-Free™, and it is designed to eliminate the effects of biofouling in the RO system. The location of the plant, as well as a bird's-eye view photo of it, can be found in Figure 16.3.

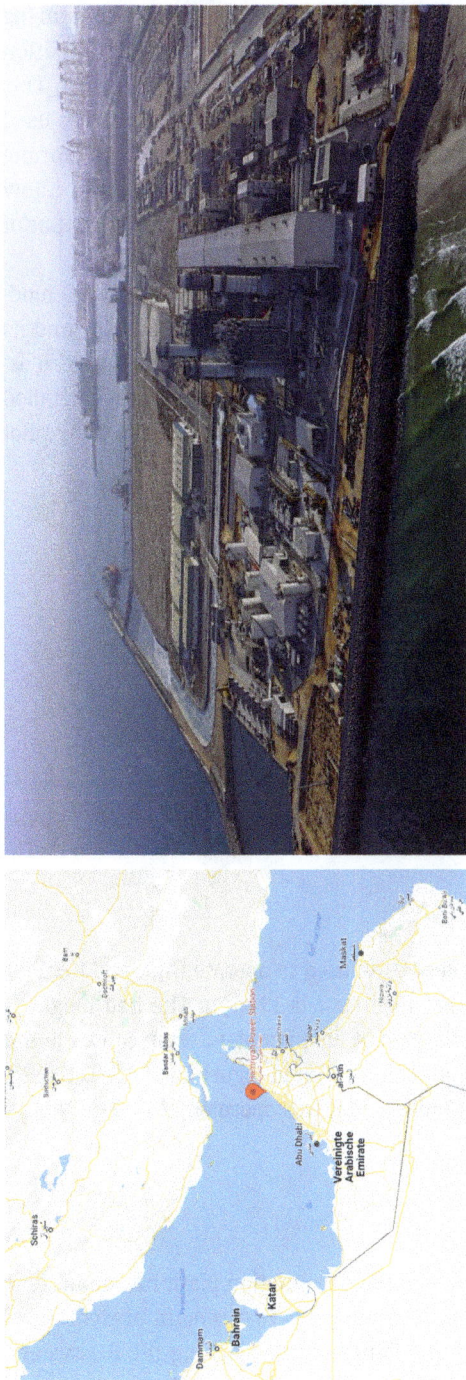

Figure 16.3: SEWA desalination plant location in the United Arab Emirates.

DuPont™ B-Free™ biofouling prevention performance was evaluated by comparing an RO element installed in the B-Free™ pilot plant to RO elements from a desalination plant as a control. A diagram of this configuration can be seen in Figure 16.4. This membrane element was cleaned through a chemical cleaning (CIP) before it was used for this side-by-side testing. The pilot plant is operated with the same RO membrane type, feed water, and operating conditions as the large desalination plant to allow comparison. In particular, the RO was operated at a feed flow of 810 m³/h, a flux of 19.5 L/m²h, and a recovery of 7.3%.

A constant dP over time indicates that biofilm is not developing within the RO module. The desalination plant was operated without DuPont™ B-Free™ technology, and its dP trend was used to quantify the number of chemical cleanings prevented and to assess the biofouling prevention achieved by the pretreatment. After operation, a detailed visual inspection of the RO was completed to confirm operational results in the pilot plant.

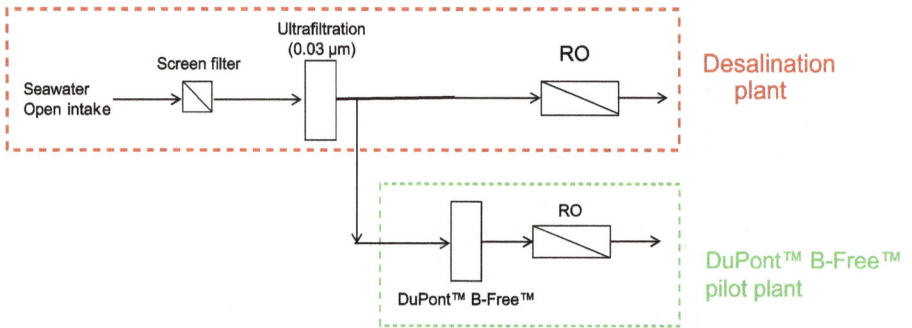

Figure 16.4: Pilot plant setup.

The glass fiber-reinforced polymer composites vessel used to operate DuPont™ B-Free™ has an inner diameter of 1,000 mm, and the feed flow is 13 m³/h. The bed height is 80 cm, corresponding to 630 L of media. A diagram of the pilot plant setup can be found in Figure 16.5. A photo of the actual pilot plant can be seen in Figure 16.6. The DuPont™ B-Free™ vessel was run at a velocity of 13–17 m/h, and it was operated at 16–21 BV/h.

16.2.2 Seawater characteristics

The open intake seawater used as feed water to the desalination plant has undergone extensive pretreatment. Because of the dissolved nutrients that remain present, however, the feedwater still has high biofouling potential once reaching the RO system, leading to biofilm build-up and a corresponding rapid dP (see Table 16.1).

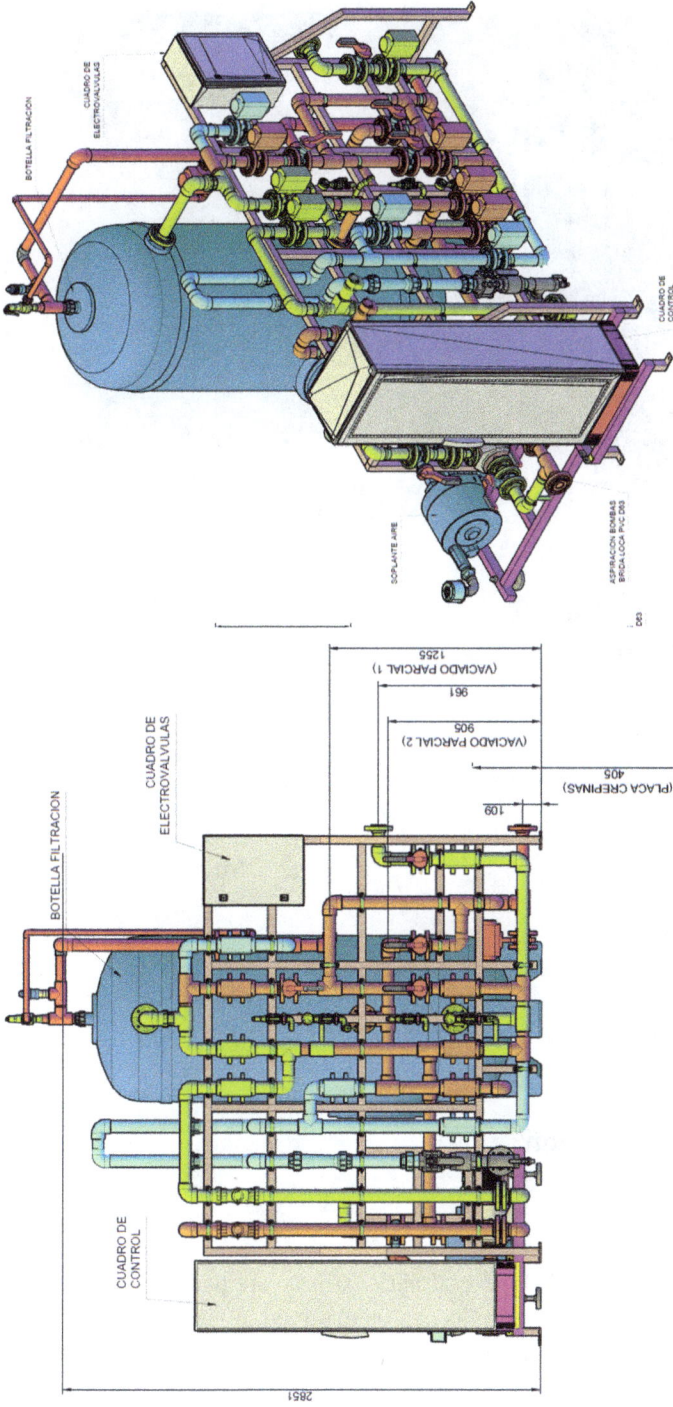

Figure 16.5: Schematic of the pilot plant.

Figure 16.6: Photo of the pilot plant.

Table 16.1: Feed water properties.

Parameter	Concentration
TDS (mg/L)	41,700
K (mg/L)	510
Na (mg/L)	13,200
Mg (mg/L)	1,300
Ca (mg/L)	400
HCO_3 (mg/L)	170
Cl (mg/L)	22,800
SO_4 (mg/L)	3,200
Boron (mg/L)	4
pH	6.8

16.3 Results and discussion

16.3.1 Backwash

During the entire operation period, several backwashes were performed to restore the dP of the DuPont™ B-Free™ system. As can be observed in Figure 16.7, the backwash protocol with water and air was effective in restoring performance without the need to use chemical cleaning.

The dP trigger used was around 0.85 bar. After the backwash, the amount of water treated before reaching the dP trigger once more can be used as a means of evaluating its effectiveness. DuPont™ B-Free™ can treat more than 2,000 m³ of raw water before a backwash is required. Considering these results and water requirements during backwashes, average water recovery has been 99.4%.

Pressure drop in the DuPont™ B-Free™ vessel

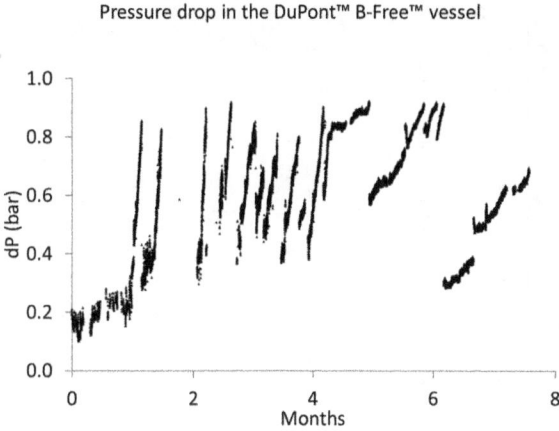

Figure 16.7: Backwash overview during operation in the DuPont™ B-Free™ vessel.

16.3.2 Reverse osmosis

Biofouling severely impacts RO operations, causing a rapid increase in dP. The desalination plant suffers from a high chemical cleaning (CIP) frequency, with an average CIP rate of 27 chemical cleanings per year.

In contrast, DuPont™ B-Free™ pretreatment in the pilot plant creates an instant and sustained biostatic environment for the downstream RO operations and is resilient to upstream upsets. In a period of eight months, an equivalent of 18 chemical cleanings would have been required in the desalination plant compared to the stable operation in the RO operated with DuPont™ B-Free™. This is equivalent to a rate of 27 chemical cleanings per year. The dP evolution over time in the RO system not protected with the biofouling prevention, as well as the RO system with the biofouling prevention technology, can be found in Figure 16.8. It should be noted that the average dP increase observed in the desalination plant that was not protected with DuPont™ B-Free™ was 50 mbar/day. On the other hand, the average dP increase in the RO protected with the DuPont™ B-Free™ pretreatment was only 0.8 mbar/day. Therefore, the DuPont™ B-Free™ prevented 18 CIPs in 8 months.

Pressure drop in RO without DuPont™ B-Free™

Pressure drop in RO with DuPont™ B-Free™

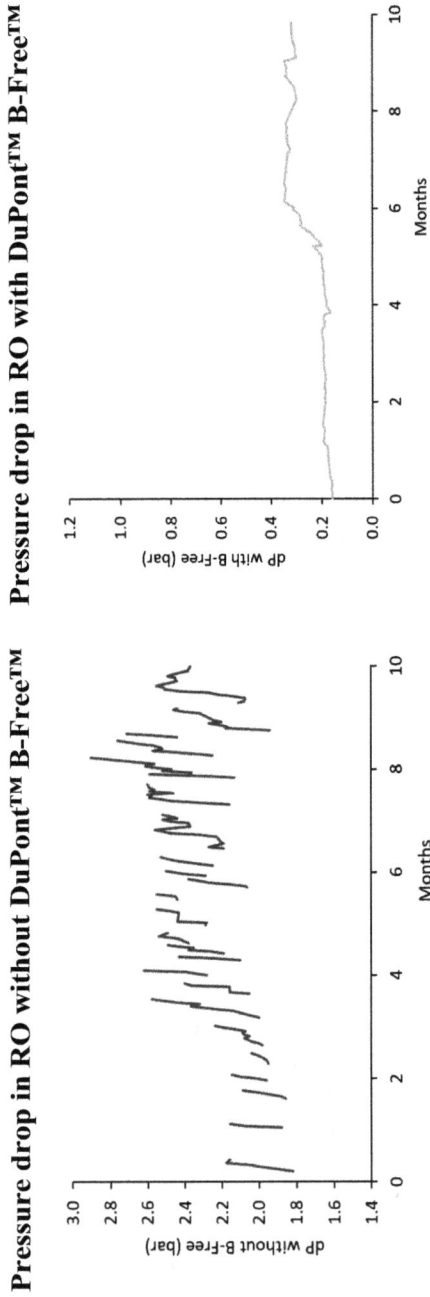

Figure 16.8: Pressure drop evolution in the reverse osmosis system not protected (left) and protected (right) with DuPont™ B-Free™ technology.

16.4 Conclusions

DuPont™ B-Free™ pretreatment technology for biofouling prevention in RO is a novel vessel-based media technology that provides instant and sustained biofouling protection. It has been validated using seawater at a commercial desalination plant with industrial-sized elements. For nearly 1 year of piloting, it provided stable operation, neutralizing the high biofouling present in the feed water.

Moreover, DuPont™ B-Free™ is compact and easy to operate, requiring only air and water for cleaning, thus eliminating the need for chemicals.

References

[1]   L. Javier, N. M. Farhat, P. Desmond, R. V. Linares, S. Bucs, J. C. Kruithof, J. S. Vrouwenvelder. Biofouling control by phosphorus limitation strongly depends on the assimilable organic carbon concentration. Water Research, 183 (2020) 116051.

[2]   T. Bhattacharjee, S. S. Datta. Confinement and activity regulate bacterial motion in porous media. Soft Matter, 15 (2019) 9920–9930.

[3]   M. Schulz, J. Winter, H. Wray, B. Barbeau, P. Bérubé. Biologically active ion exchange (BIEX) for NOM removal and membrane fouling prevention. Water Science and Technology: Water Supply, 17 (2017) 1178–1184.

[4]   H. Yang, B. Minuth, D. Grant Allen. Effects of nitrogen and oxygen on biofilter performance. Journal of the Air & Waste Management Association, 52 (2002) 279–286.

[5]   H. C. Flemming. Biofouling in water systems – Cases, causes and countermeasures. Applied Microbiology and Biotechnology, 76(6) (2007) 1255–1267.

[6]   L. Vanysacker, R. Bernshtein, I. F. J. Vankelecom. Effect of chemical cleaning and membrane aging on membrane biofouling using model organisms with increasing complexity. Journal of Membrane Science, 457 (2014) 19–28.

[7]   Z. Liu, K. M. Lompe, M. Mohseni, P. R. Bérubé, S. Sauvé, B. Barbeau. Biological ion exchange as an alternative to biological activated carbon for drinking water treatment. Water Research, 168 (2020) 115148.

[8]   S. West, H. Horn, W. A. M. Hijnen, C. Castillo, M. Wagner. Confocal laser scanning microscopy as a tool to validate the efficiency of membrane cleaning procedures to remove biofilms. Separation and Purification Technology, 122 (2014) 402–411.

[9]   H. Wray, M. Schulz, J. Winter, P. R. Bérubé, B. Barbeau, W. Bayless. Biological ion exchange for NOM removal – unforeseen synergies. American Water Works Association – Water Quality Technology Conference (WQTC), (2016) 13–17.

[10]   M. Clements, J. Haarhoff. Practical experiences with granular activated carbon (GAC) at the rietvlei water treatment plant. Water SA, 30 (2004) 89–95.

[11]   T. Nguyen, F. A. Roddick, L. Fan. Biofouling of water treatment membranes: A review of the underlying causes, monitoring techniques and control measures. Membranes, 2 (2012) 804–840.

[12]   B. D. Martin, L. De Kock, M. Gallot, E. Guery, S. Stanowski, J. MacAdam, B. Jefferson. Quantifying the performance of a hybrid anion exchanger/adsorbent for phosphorus removal using mass spectrometry coupled with batch kinetic trials. Environmental Technology, 39 (2018) 2304–2314.

[13] O. Lorain, B. Hersant, F. Persin, A. Grasmick, N. Brunard, J. M. Espenan. Ultrafiltration membrane pre-treatment benefits for reverse osmosis process in seawater desalting. Quantification in terms of capital investment cost and operating cost reduction. Desalination, 203(1–3) (2007) 277–285.

[14] F. Fatima, S. Fatima, H. Du, R. R. Kommalapati. An evaluation of microfiltration and ultrafiltration pretreatment on the performance of reverse osmosis for recycling poultry slaughterhouse wastewater. Separations, 11(4) (2024) 115.

Index

https://doi.org/10.1515/9783111639079-017

Back cover

Have you ever wondered how biofouling takes hold in spiral-wound reverse osmosis or nanofiltration membranes? Despite decades of advancements, biofouling remains one of the most persistent challenges in the water treatment industry.

This book tackles these critical questions head-on:

– How does temperature influence biofouling development?
– What role can biocides play in halting biofouling?
– How can biofouling be effectively studied and understood?
– Which parameters are most crucial for biofouling prevention?
– How can biofouling be effectively prevented?

Through a blend of scientific rigor and practical insights, this book offers a comprehensive guide to understanding, mitigating, and preventing biofouling and organic fouling. Whether you are a researcher, engineer, or water treatment professional, you will gain valuable strategies and evidence-based approaches to address one of the water treatment industry's most pressing issues.

https://doi.org/10.1515/9783111639079-018

www.ingramcontent.com/pod-product-compliance
Lightning Source LLC
Chambersburg PA
CBHW061345210326
41598CB00035B/5883